Scientific Knowledge Series

NEET – UG / ICAR AIEEA – UG / TNPSC A.O

BIOLOGICAL SCIENCE

Authors

R. Thangapandian

A. Yuvaraja

SHANLAX PUBLICATIONS

BIOLOGICAL SCIENCE

First Edition: November 2020

ISBN: 978-93-90082-77-3

Publishers' Note

The present work is a humble attempt to satisfy the needs of students appearing for NEET- UG / ICAR AIEEA - UG / A.O Examinations. After theory and MCQs, model papers are provided for practice. The book in the present form shall serve as a reliable evocative book for those preparing for Under graduation in medicine / Agriculture / Agricultural Officers Examinations

Any suggestions for the improvement shall be thankfully acknowledged.

--- with Best Wishes

Recommended for Eligibility for NEET - UG / ICAR AIEEA -UG / Agricultural Officer in Banks and State Government Level and UGC / CSIR JRF - SRF Examinations

Publisher
SHANLAX PUBLICATIONS
61, 66 T.P.K. Main Road
Vasantha Nagar
Madurai – 625003
Tamil Nadu, India

Ph: 0452-4208765,
Mobile: 7639303383
email:publisher@shanlaxpublications.com
web: www.shanlaxpublications.com

CONTENTS

CHAPTER - I
CROP BOTANY

Symbol of sex

The symbol for the male ♂ represents the shield and spear of the Greek god of war, Zeus, and the female symbol ♀ represents the vanity mirror of the goddess of love, Aphrodite.

Position of Ovary

1. Hypogynous - Superior ovary (eg) citrus, brinjal
2. Perigynous - semi inferior ovary (eg) Pea
3. Epigynous - inferior ovary (eg) Apple, sunflower, Guava, Coriander

Pollination

1. Autogamy - self pollination
2. Allogamy - cross pollination
3. Geintonogamy - deposition of pollen on stigma of another flower of the same inflorescence.
4. Ormithophily - birds pollination
5. Entomophily - insect pollination

- Which one is the non - endospermic seed?
- a) Maize **b) Pea** c) Rice d) Wheat
- Ginger is a stem and not a root because it has nodes and internodes.
- Leaves fall of from branches in winter on account of formation of abscission layer.
- Insectivorous plants grow in soil deficient in Nitrogen.
- Balanophora species are total root parasites.

- The plant whose seeds are known to have longest viability period: Lotus (*Nelumbo nucifera*). (eg) 400 years (viability) dormancy.
- Rice differs among cereal crops because it has 6 stamens.
- The longest stigma is found in Maize.
- Till today, largest flower in the world is that of Rafflesia.
- Grafting can be successful if there is fusion of cambium.

Types of Fruits

S. No.	Type of fruits	Crops
A.	**Simple fruit:**	
	1. Berry	Banana, Papaya, Grape, Sapota, Avocada, Tomato
B.	**Modified berry**	
	1. Balusta	Pomegranate
	2. Amphisarca	Woodapple, Bael
	3. Pepo	Water melon
	4. Pome	Apple, Pear, Loquat
	5. Drupe (Stone)	Mango, Pear, Plum, Coconut
	6. Hesperidium	Citrus
	7. Nut	Cashew, Litchi, Walnut, Rambutan
	8. Capsule	Anola, Carambola
C.	**Aggregate Fruits**	
	1. Etario of berries	Custard apple, Raspberry
D.	**Multiple fruit**	
	1. Syconus	Fig
	2. Sorosis	Jack fruit, Pineapple, Mulberry

- The apple fruit is categorized as false because its actual fruit is located within edible flashy thalamus.
- In fig the edible part is juicy peduncle
- If seed is defined as an ovule, modified as a result of fertilization, one may except to final seeds in phanerogams only.
- (Phanerogams: includes gymnosperms and angiosperms)
- In Bougainvillea, the large coloured structures are : Petaloid bracts
- A monocot flower differ from the dicot flower in, monocot flowers have perianth.
- The juicy edible portion of cashew nut is pedicel.

Edible part of fruits

1. Orange : multicelled endocarpic hairs
2. Pomegranate : Juicy testa
3. Grape : Pericarp and placenta
4. Cucumber : pericarp, placenta and seed
5. Cashewnut : Pedicel
6. Apple : Thalamus
7. Banana : mesocarp and endocarp

- Sudden increase in the respiration during fruit ripening is called climacteric.
- **Monellin** is a sweet - tasting protein found in some fruits, when it is denatured, it no longer tastes sweet.
- The first amino acid isolated from a protein hydrolyzate was glycine, obtained in 1820 from gelatin by H. Braconnot.

Crop wise Geographical Indication in India

S. No.	Geographical indication	Year	State
1.	Darjeeling Tea (word & logo)	2005	West Bengal
2.	Kangra Tea	2006	Himachal Pradesh
3.	Coorg Orange	2006	Karnataka
4.	Mysore Betel leaf	2007	Karnataka
5.	Nanjanagud Banana	2007	Karnataka
6.	Mysore Jasmine	2008	Karnataka
7.	Udupi Jasmine	2008	Karnataka
8.	Hadagali Jasmine	2008	Karnataka
9.	Navara Rice	2008	Kerala
10.	PalakkadanMatta Rice	2008	Kerala
11.	Malabar Pepper	2008	Kerala
12.	Allahabad Surkha	2008	Uttar Pradesh
13.	Monsooned Malabar Arabica Coffee	2008	Karnataka
14.	Monsooned Malabar Robusta Coffee	2008	Karnataka
15.	Spices -Alleppey Green Cardamom	2008	Kerala
16.	Coorg Green Cardamom	2008	Karnataka
17.	Eathomozhy Tall Coconut	2009	Tamil Nadu
18.	Pokkali Rice	2009	Kerala
19.	LaxmanBhog Mango	2009	West Bengal
20.	Khirsapati (Himsagar) Mango	2009	West Bengal
21.	Fazli Mango grown in the district of Malda	2009	West Bengal
22.	Naga Mircha	2009	Nagaland

23.	Nilgiri (Orthodox) Logo	2009	Tamil Nadu
24.	Assam (Orthodox) Logo	2009	Assam
25.	Virupakshi Hill Banana	2009	Tamil Nadu
26.	Sirumalai Hill Banana	2009	Tamil Nadu
27.	Mongo MalihabadiDusseheri	2010	Uttar Pradesh
28.	Vazhakulam Pineapple	2010	Kerala
29.	DevanahalliPomello	2010	Karnataka
30.	Appemidi Mango	2010	Karnataka
31.	Kamalapur Red Banana	2010	Karnataka
32.	Bikaneri Bhujia	2011	Rajasthan
33.	Guntur Sannam Chilli	2011	Andhra Pradesh
34.	Mahabaleshwar Strawberry	2011	Maharashtra
35.	WayanadJeerakasala Rice	2011	Kerala
36.	WayanadGandhakasala Rice	2011	Kerala
37.	Central Travancore Jaggery	2011	Kerala
38.	Nashik Grapes	2011	Maharashtra
39.	Byadagi chilli	2011	Karnataka
40.	GirKesar Mango	2012	Gujarat
41.	Bhalia Wheat	2012	Gujarat
42.	UdupiMattuGullaBrinijal	2012	Karnataka
43.	GanjamKewdaRooh	2012	Orissa
44.	Ganjumkewda Flower	2012	Orissa
45.	Madurai Malli	2013	Tamil Nadu
46.	Bangalore Blue Grapes	2013	Karnataka
47.	Kalanamak Rice	2013	Uttar Pradesh

Specialties in plant kingdom

1.	Tallest living tree (gymnosperm)	:	*Sequoia sempervirens* (Red wood tree, Father of the forest)
2.	Tallest angiosperm	:	*Eucalyptus regnans* more than 80 meter height

3.	Largest cactus	:	*Carnegia giganteum*
4.	Tallest pteridophyte	:	Alsophila
5.	Largest moss	:	Dawsonia
6.	Largest sea weed (alga)	:	*Macrocystis pyrifera*
7.	Smallest angiosperm	:	*Wolffia angusta*
8.	Smallest gymnosperm	:	*Zamia pygmaea*
9.	Smallest pteridophyte	:	Azolla
10.	Smallest angiospermic parasite	:	Arceuthobium
11.	Largest vegetative bud	:	Cabbage
12.	Longest compound leaf	:	*Raphia taedigera*
13.	Largest simple leaf	:	*Victoria regia*
14.	Largest inflorescence	:	*Puya raimondii*
15.	Smallest inflorescence	:	Lemna
16.	Largest flower	:	*Rafflesia arnoldi*
17.	Smallest flower	:	Wolffia
18.	Largest fruit	:	*Citrullus lanatus*
19.	Largest ovule	:	Cycas
20.	Largest egg	:	Cycas
21.	Largest spermatozoids	:	Cycas
22.	Longest style	:	Maize
23.	Largest seeds	:	*Lodoicea maldivica*
24.	Smallest seeds	:	Orchids
25.	Heaviest wood	:	*Olea laurifolia*
26.	Lightest wood	:	*Aeschynomene hispida*
27.	Most viable seeds	:	*Lupinus arcticus*
28.	Biggest antherozoid	:	Cycas
29.	Largest angiospermic family	:	Asteraceae

30.	Largest Botanical garden in the world	:	The Royal Botanic garden, Kew, England
31.	The largest Botanical garden in India	:	Indian Botanic garden, Calcutta
32.	Largest herbarium in the world	:	The Royal Botanic garden, Kew, England
33.	Largest herbarium in India	:	Central National herbarium, Sibpur, Howrah
34.	Religious tree of Hindus	:	*Ficus religiosa*
35.	State tree of Tamil Nadu	:	Palmyra
36.	State tree of Andhra Pradesh	:	*Azaderachta indica* (Neem)
37.	National fruit of India	:	Mango
38.	National flower of India	:	Lotus (*Nelumbo nucifera*)
39.	State flower of TamilNadu	:	*Gloriosa superpa* (Karthikai poo)
40.	The largest plant cell	:	Acetabularia
41.	Smallest plant cell	:	*Dialister pneumosintes*
42.	Longest fibre	:	Ramie (*Boehmeria nivea*)
43.	Longest vessel	:	Eucalyptus
44.	Least no. of chromosomes	:	*Haplopappus gracilis*
45.	Highest no. of chromosomes	:	*Morus nigra*
46.	Longest plant chromosomes	:	Trillium
47.	Angiospermic saprophyte	:	Monotropa
48.	Most poisonous fungus	:	*Amanita phalloides*

49.	Enucleate plant cells	:	Sieve cells, Sieve tube elements
50.	The most primitive flower	:	Magnolian flower

Books and Authors

1.	Historia plantarum	:	Theophrastus
2.	Species plantarum	:	Linnaeus
3.	Systema Naturae	:	Linnaeus
4.	Micrographia	:	Robert Hooke
5.	Genera plantarum	:	Bentham & Hooker
6.	Die Naturlichen Pflanzenfamilien	:	Engler & Pranti
7.	Families of Flowering Plants	:	John Hutchinson
8.	The evolutionary tendencies in flowering plants	:	Bessey
9.	The Dictionary of Flowering Plants & Ferns	:	Willis
10.	Anatomia Plantarum	:	Marcello Malpighi
11.	Origin of Species	:	Charles Darwin
12.	Flora of British India	:	J.D.Hooker
13.	Flora of Madras Presidency	:	G.S.Gamble
14.	Manual of Determinative Bacteriology	:	Bergey
15.	Structure & Reproduction of Algae	:	F.E.Fritsch
16.	Introductory Mycology	:	Alexopoulos
17.	The Physical basis of Life	:	Huxley
18.	The evolution and classification of flowering plants	:	Arthur Cronquist
19.	Hortus Uplandicus	:	Linnaeus
20.	De Materia Medica	:	Dioscoroides

21.	A system of physiology of the flowering plants	:	Takhtajan
22.	Cellular Respiration	:	Meldrum

Theories & their proponents

1.	Cell theory	:	M.J. Schleiden Theodore Schwann
2.	Cell lineage theory	:	Rudolf Virchow
3.	Unit membrane concept	:	Robertson
4.	Protoplasm theory	:	Max Schultze
5.	Chromosome theory of inheritance	:	W.S. Sutton, Theodore Boveri
6.	Gene theory	:	T.H.Morgan
7.	Germplasm theory	:	Weismann
8.	Theory of pangenesis	:	Charles Darwin
9.	Mutation theory	:	Hugo de Vries
10.	Natural selection	:	Charles Darwin
11.	Inheritance of acquired characters	:	Lamark
12.	Apical cell theory	:	Hofmeister
13.	Histogen theory	:	Hanstein
14.	Tunica-Corpus theory	:	Schmidt
15.	Telome theory	:	Zimmerman
16.	Stelar theory	:	Van Tiegham & Douliot
17.	Theory of Organic evolution	:	Charles Darwin
18.	Gene Concept	:	S.Benzer
19.	Laws of heredity	:	G.J.Mendel
20.	Operon concept	:	Jacob & Monod
21.	One gene-one enzyme hypothesis	:	Beadle & Tatum
22.	One gene-one polypeptide hypothesis	:	V.M.Ingram

23.	Ion exchange theory	:	Jenney & Overstreet
24.	Donnons equilibrium theory	:	F.G. Donnon
25.	Carrier concept	:	Van den Honert
26.	Pulsation theory	:	J.C. Bose
27.	Root pressure theory	:	Stephen Hales
28.	Cohesion-Tension theory	:	Dixon & Jolly
29.	Starch-sugar hypothesis	:	Lloyd
30.	Lock & Key mechanism	:	Emil Fisher
31.	Induced fit theory	:	Koshland
32.	Mass flow or Pressure flow hypothesis	:	E.Munch
33.	Law of minimum	:	Liebig
34.	Law of limiting factors	:	F.F.Blackmann
35.	Chemiosmotic theory	:	Mitchel
36.	Cholodney-Went hypothesis	:	Cholodny & F.W.Went
37.	Sex linked inheritance	:	T.H.Morgan
38.	Criteria of essentiality of elements	:	D.I.Arnon and Stout
39.	Two pigment system	:	Emerson
40.	Red drop and Emerson effect	:	Emerson *et al*
41.	Photophosphorylation	:	D.I.Arnon

Fathers of various faculties

1.	Father of Biology	:	Aristotle
2.	Father of Botany	:	Theophrastus
3.	Father of Modern Botany	:	Linnaeus
4.	Father of Phycology	:	G.Kelbs
5.	Father of Mycology	:	Antonio Micheli

6.	Father of Modern mycology	:	Anton de Bary
7.	Father of Microbiology	:	Leeuwenhoek
8.	Father of Bacteriology	:	Leeuwenhoek
9.	Father of Virology	:	Robert Mayer
10.	Father of Lichenology	:	Ahmadjian
11.	Father of immunology	:	Edward Jenner
12.	Father of Enzymology	:	Edward Buchner
13.	Father of cytology	:	Robert Hooke
14.	Father of Genetics	:	Gregor Johan Mendel
15.	Father of Modern Genetics	:	T.H.Morgan
16.	Father of Evolution	:	Charles Darwin
17.	Father of Plant Breeding	:	Thomas Fairchild
18.	Father of Plant Taxonomy	:	Carolus Linnaeus
19.	Father of Physiology	:	Stephen Hales
20.	Father of Embryology	:	C.F.Wolff
21.	Father of Tissue culture	:	Haberlandt
22.	Father of Green revolution	:	Norman E. Borlaug
23.	Father of Epidemology	:	John Snow
24.	Father of Horticulture	:	John Lindley
25.	Father of Hybrid Rice	:	Yuan Long Ping
26.	Father of Indian Botany	:	Dr. William Roxburgh

Founders of techniques

1.	Chromatography	:	M.Twett
2.	Isotope tracing technique or Autoradiography	:	G.Hevesy
3.	Feulgen technique	:	Feulgen & Rossenbeck
4.	Ultracentrifugation/Cell fractionation	:	Svedberg
5.	Iodine test	:	F.V.Raspail
6.	Tissue culture technique	:	Haberlandt

7.	Microtomy	:	W.His
8.	X-ray diffraction & Crystallography	:	Rosalind Franklin & Wilkins
9.	UV Photomicrography	:	T.Casperson
10.	Ringing experiment	:	Malpighi

Experiments

1.	Thistle funnel experiment	:	To demonstrate Osmosis
2.	Ringing experiment	:	Ascent of sap, Translocation of organic solutes
3.	Porous pot experiment	:	To demonstrate Ascent of sap (Cohesion-Tension theory)
4.	Bell jar experiment	:	To demonstrate Transpiration
5.	Cobalt chloride experiment	:	To compare the rates of Transpiration
6.	Hydroponic experiment	:	To prove the essentiality of the element
7.	Hydrilla-Funnel experiment	:	To demonstrate the evolution of O_2 during photosynthesis
8.	Molls half leaf experiment	:	To demonstrate that CO_2 is essential for photosynthesis
9.	Ganongs light screen experiment	:	To demonstrate that light is essential for photosynthesis

10.	Variegated leaf experiment	:	To demonstrate that chlorophyll is essential for photosynthesis
11.	Dewars flask experiment	:	To demonstrate the temperature increase during respiration
12.	Heliotropic chamber experiment	:	To demonstrate phototropism
13.	3 bean seed experiment	:	To demonstrate the essentiality of water & O_2 in seed germination

Important Discoveries

1.	Cell	:	Robert Hooke (1665)
2.	Free living cell	:	Anton Von Leeuwenhoek
3.	Cell wall	:	Robert Hooke
4.	Plasma membrane	:	W.Pfeffer
5.	Nucleus	:	Robert Brown
6.	Nucleolus	:	Fontana
7.	Endoplasmic reticulum	:	K.R.Porter (1945)
8.	Microsomes	:	Claude
9.	Ribosomes	:	Palade
10.	Golgi complex	:	Camello Golgi
11.	Chloroplast	:	Sachs
12.	Mitochondria	:	Kolliker
13.	Mesosomes	:	Fitz-James
14.	Oxisomes	:	Lehninger
15.	Cytochromes	:	Mc Munn
16.	ATP	:	Lohmann
17.	Lysosomes	:	Christian de Duve

18.	Peroxisomes	:	Tolbert
19.	Glyoxysomes	:	Beevers
20.	Sphaerosomes	:	Hanstein
21.	Microtubules	:	Ledbetter and Porter
22.	Quantasome	:	Park & Biggins
23.	Cristae	:	Palade
24.	Chromosome	:	Hofmeister
25.	Giant chromosome (polytene chromosome)	:	Balbiani
26.	Lamp brush chromosome	:	W.Flemming
27.	Sex or X-chromosome	:	Mc Clung
28.	Nucleic acid	:	Friedrich Miescher
29.	mRNA	:	Jacob & Monod
30.	tRNA	:	P.C.Zamecnic, Hoagland
31.	Plasmid	:	Jacob & Wollman
32.	DNA polymerase	:	Aurther Kornberg
33.	Polynucleotide phosphorylase	:	S.Ochoa & Grunberg Manago
34.	Bacteria	:	Anton Von Leuwenhoek
35.	Viruses	:	Iwanowski
36.	Bacteriophage	:	Twort & d' Herelle
37.	Cyanophage	:	Safferman & Moris
38.	Penicillin	:	Alexander Fleming
39.	Streptomycin	:	Waksman
40.	Enzymes	:	Buchner
41.	Amitosis	:	Robert Remak
42.	Mitosis	:	Strasburger
43.	Auxin	:	Kogl-Haagen Smit-Erx leben

44.	Gibberellin	:	Yabuta-Hayashi
45.	Cytokinin	:	Skoog & Miller
46.	Abscissic acid	:	Eagle & Wareing
47.	Co.A	:	F.A.Lipman
48.	Transposans (Jumping gene)	:	Mc.Clintock
49.	Mutations	:	Hugo de Vries
50.	Photoperiodism	:	Garner & Allard
51.	Induced Mutation	:	H.J. Muller
52.	Sieve tube	:	Hartig
53.	Operon	:	Jacob and Monod
54.	Laticiferous tubes	:	De Bary
55.	Poly embryony	:	Anton Von Leeuwenhoek
56.	Heterothallism	:	Blakeslea
57.	Double fertilization	:	S.G.Nawaschin
58.	Syngamy	:	Strasburger
59.	Cytoplasmic inheritance	:	J.M.Sonneborn
60.	Bacterial transformation	:	Fred Griffith
61.	Bacterial conjugation	:	Lederberg & Tatum
62.	Transduction	:	Zinder & Lederberg
63.	Genetic code	:	Nirenberg *et al*; Khorana *et al*

Important terms coined

1.	Cell	:	Robert Hooke
2.	Protoplast	:	Hanstein
3.	Cellular totipotency	:	Steward

4.	Prokaryote/Eukaryote	:	Dougherthy
5.	Protoplasm	:	Purkinjie
6.	Cytoplasm	:	Kolliker
7.	Entoplasmic reticulum	:	K.R.Porter
8.	Ribosomes	:	Palade
9.	Bioblast	:	Altmann
10.	Mitochondrion	:	Benda
11.	Plastid	:	Haeckel
12.	Chloroplast	:	Schimper
13.	Thylakoid	:	Menke
14.	Quantasome	:	Park & Biggins
15.	Microsome	:	A.Claude
16.	Chromosome	:	Waldeyer
17.	Nuclein	:	F.Miescher
18.	Nucleic acid	:	Altmann
19.	Chromatin	:	Fleming
20.	Euchromatin & heterochromatin	:	Heitz
21.	Sex chromosome	:	Wilson
22.	Nucleosome	:	Woodcock
23.	Lamp brush chromosome	:	Ruckert
24.	Cistron, Recon, Muton	:	S. Benzer
25.	mRNA	:	Jacob & Monod
26.	Mitosis & Amitosis	:	Walter Fleming
27.	Meiosis	:	Farmer & Moors
28.	Meristem	:	C.Nageli
29.	Parenchyma	:	Nehemiah Grew

30.	Collenchyma	:	Schleiden
31.	Sclerenchyma	:	Mettenius
32.	Sclereid	:	Tschirch
33.	Xylem & Phloem	:	C.Nageli
34.	Tracheid	:	Sanio
35.	Sieve tube	:	Hartig
36.	Protein	:	G.J.Mulder
37.	Vitamin	:	Cassimir Funk
38.	Enzyme	:	Kuhne
39.	Auxin	:	Kogi
40.	Gibberellin	:	Yabuta
41.	Cytokinin	:	Letham
42.	Parthenocarpy	:	Noll
43.	Polyembryony	:	Leeuwenhoek
44.	Protista	:	Haeckel
45.	Genetics	:	Bateson
46.	Mutation	:	Hugo de Vries
47.	Ecology	:	Reiter
48.	Ecosystem	:	A.G.Tarlsley
49.	Taxonomy	:	A.P. de Candolle

Derivatives of Plant names

Nicotiana: after **Nicot**, a French Ambassador in Portugal, who obtained seeds of tobacco.

Largest inflorescence: *Puya raimondii*

Puya raimondii's inflorescence can grow up to 30 feet, with more than 30 thousand flowers and 6 million seeds; after flowering it dies. This plant is found in the higher elevations of Peru (around 12,000 feet).

Tallest living tree (angiosperm)

The world's tallest angiosperm tree is *Eucalyptus regnans* in wet forests of Tasmania, Australia. It is commonly called as **Mountain Ash, Victorian Ash, Swamp Gum, Tasmanian Oak** or **Stringy Gum.** They grow very quickly, at more than a metre a year, and can reach 65 metres (213 ft) in 50 years, and it has been known to attain heights over **100 metres** (330 ft) with an average life-span of **400 years**. It is propagated only from seed. The seeds are released from their woody capsules (gum nuts) by heat and for successful germination the seedlings require a high level of light. Primary uses are furniture, flooring, panelling, veneer, plywood, window frames and general construction.

Tallest living tree (gymnosperm)

Sequoia sempervirens (Father of the forest), its common names are **coast redwood**, **giant redwood** and **California redwood**. The tallest trees on Earth, reaching up to **115.61 metres** (379.30 ft) in height and up to 7.9 metres (25.9 ft) diameter at breast height. It is an evergreen, long-lived, living for up to **2,500 – 3,500 years** or more. The species is monoecious, with pollen and seed cones on the same plant. Its genetic makeup is unusual among conifers, being a hexaploid (6n) and possibly allopolyploid (AAAABB). The mitochondrial genome is (unlike other conifers) paternally inherited. Coast redwood reproduces both sexually and asexually. Seed production begins at 10–15 years of age, but viability of the seed is low (<15%). Timber is highly valued for its beauty, light weight, and resistance to decay. **Its lack of resin makes it resistant to fire**. Because of its impressive resistance to decay, redwood was extensively used for rail road ties and trestles throughout California

Largest simple leaf: ***Victoria regia***

Victoria is a genus of water lilies, in the plant family Nymphaeaceae, with very large leaves that float on the water's surface. It has a leaf that is up to **3 m (9.8 ft) in diameter**, on a **stalk 7–8 m (22.9-26.2 ft) in length**. The genus name was given in honour of **Queen Victoria** of the United Kingdom.

Largest flower: ***Rafflesia arnoldi***

Several species of Rafflesia grow in the jungles of Southeast Asia, all of them are endangered. It produces no leaves, stems or roots but lives as a parasite on the *Tetrastigma* vine, which grows only in primary (undisturbed) rainforest. Only the flower or bud can be seen; the rest of the plant exists only as filaments within its unfortunate host. The blossom is pollinated by flies attracted by its scent, which resembles that of carrion.

The world's largest flower can have a **diameter up to one meter** and can **weigh up to 10 kilograms**. It also smells like rotting flesh. Stamens and pistils fused together in a central column, and both produce a corona, or crown, in the shape of a ring. The flowers bloom only once a year and live for five to seven days. 20 species of genus Rafflesia is reported.

Longest compound leaf: ***Raphia taedigera***

This plant holds the world record for "longest leaves" for its leaves that can reach **65 feet long**. The Raphia palms are a genus of twenty species of palms native to tropical regions of Africa, especially Madagascar, *Raphia taedigera* also occurring in Central and South America. They grow up to **16 m tall** and are remarkable for their compound pinnate leaves, the longest in the plant kingdom.

Least chromosome number: *Haplopappus gracilis*

Haplopappus gracilis, has the lowest chromosome number (2n = 4) belonged to the Compositae (Sunflower) family, annual plants, flowers with yellow rays and disks to 1.25" wide. Leaves simple, toothed with spiny tip to each tooth. Flowers bloom from February to December.

Highest chromosome number: Black Mulberry

Black Mulberry (***Morus nigra***) possesses the highest chromosome number (2n=308). It has berry-like fruit with a unique tart-sweet taste. The black mulberry is generally considered the best tasting of the various mulberry species. Medium sized deciduous tree usually only to 20-30 feet, but can reach 70 feet. Trees are known to be very long and can bear fruit for several hundred years. The propagation is usually by seed, which can take several years to bear. It is native to western Asia. It has been cultivated in Europe for hundreds of years.

Crop Improvement

- The term mutation was first used by Hugo de Vries to describe the sudden phenotypic changes which were heritable, while working with *Oenothera lamarckiana*.
- But the earliest record of mutations dates back to 1791 when Seth Wright noticed a male lamb with unusually short legs in his flock of sheep. This lamb served as a source of short leg trait for the development of Ancon breed of sheep.
- However the systematic studies on mutations were started in 1910 by T.H.Morgan who used *Drosophila melanogaster* for his studies.
- In 1927, H.J. Muller demonstrated for the first time the artificial induction of mutations by using x-rays in

Drosophila. Similarly in 1928, L.J. Stadler demonstrated an increase in the rate of mutations due to x-rays in barley and maize.

- Induction of mutations by chemicals in fungus *Aspergillus* was demonstrated by R.A. Steinberg in 1939.
- Auerbach and J.N. Robson in 1946 used chemicals to induce mutations in *Drosophila*.
- The first plant breeding programme using mutations (mutation breeding) was initiated in 1929 in Sweden, Germany and Russia. In India it was initiated in early 1930s.

In crop plants some varieties were developed through spontaneous mutations. They are

1. Rice: GEB-24, Dee-Geo-Woo-Gen
2. Wheat: Norin
3. Groundnut: TMV-10
4. Sorghum: Co-4 (Coimbatore 4)

- **Induced mutations:** Mutations can be induced artificially through treatment with either physical or chemical mutagens. The exploitation of induced mutations for crop improvement is called mutation breeding. The rate of induced mutations is very high. The induced mutations did not differ from spontaneous mutations in expression.

Examples of popular induced mutants in crop plants are:

S. No	Crop	Mutant variety	Original variety	mutagen
1.	Rice	Jagannath	T-141	X-rays
		Mahsuri mutant	Mahsuri,	g-rays
2.	Wheat	Sharbati sonara	Sonara 64	UV rays
		NP-836	NP-799	x rays
3.	French	Pusa Parvati	Wax podded	x-rays

	Beans	Pusa Lal Meeruti	Meeruti	x-rays
4.	Tomato	S-12	Sioux	g-rays
5.	Castor	Aruna	HC-6	Thermal neutrons
6.	Cotton	MCU 7	1143 EE	x-rays
		MCU 10	MCU 4	g-rays

Plant introduction agencies in India

1. Botanical Survey of India (BSI)
2. Indian Council of Forestry Research and Education (ICFRE)
3. National Bureau of Plant Genetic Resources (NBPGR)

Chromosome Numbers of Selected Plant species

S. No.	Common Name	Binominal Name	Chromosome number (2n)
1.	Black Mulberry	*Morus nigra*	**308**
2.	Adder's tongue fern	*Ophioglossum vulgatum*	**240**
3.	Willow, Grey-leaf	*Salix glauca*	**152**
4.	Cord grass	*Spartina townsendii*	**126**
5.	Saltmarsh grass	*Spartina alterniflora*	**28, 40, 42, 60, 62, 84, 120, 122, 124**
6.	Sweet Potato	*Ipomoea batates*	**90**
7.	Coffee	*Coffea arabica*	**22, 44, 66, 88**
8.	Banana	*Musa paradisiaca*	**22,44,55,77, 88**
9.	Sugar cane	*Sacharum officinarum*	**80**
10.	Spartina	*Spartina alternifolia*	**70**
11.	Pear	*Pyrus communis*	**34,51,68**
12.	Redwood, Coast	*Sequoia sempervirens*	**66**
13.	Dahlia, cultivated	*Dahlia variabilis*	**64**
14.	Orchid	*Miltoniopsis*	**56**
15.	Cloudberry	*Rubus chamaemorus*	**56**
16.	Azolla	*Azolla nilotica*	**52**

17.	Cotton, Upland	*Gossypium hirsutum*	**52**
18.	Tobacco	*Nicotiana tabaccum*	**48**
19.	Plum	*Prunus domestica*	**48**
20.	Potato	*Solanum tuberosum*	**48**
21.	Autumn Crocus	*Colchium Autumnale*	**38,42**
22.	Oat	*Avena sativa* L.	**42**
23.	Wheat	*Triticum aestivum*	**42**
24.	Peanut	*Arachis hypogae*	**40**
25.	Orange	*Citrus sinensis*	**18,27,36**
26.	Sunflower	*Helianthus annuus*	**34**
27.	Cherry	*Prunus cerasus*	**32**
28.	Rose	*Rosa*	**14, 28**
29.	Tomato	*Solanum lycopersicum*	**24**
30.	Rice	*Oryza sativa*	**24**
31.	Bean	*Phaseolus vulgaris*	**22**
32.	Cowpea	*Vigna unguiculata*	**22**
33.	Sorghum	*Sorghum vulgare*	**20**
34.	Hop	*Humulus lupulus*	**18, 20**
35.	Broccoli	*Brassica oleracea*	**18**
36.	Cabbage	*Brassica oleracea*	**18**
37.	Beet	*Beta vulgaris*	**18**
38.	Radish	*Raphanus sativus*	**18**
39.	Carrot	*Daucus carota*	**18**
40.	Black Mustard	*Brassica nigra*	**16**
41.	Onion	*Allium cepa*	**16**
42.	Snapdragon	*Antirrhinum majus*	**16**
43.	Pea	*Pisum sativum*	**14**
44.	Evening Primrose	*Oenothera lamarckiana*	**14**
45.	Sweet pea	*Lathyrus odoratus*	**14**
46.	Cucumber	*Cucumis sativus*	**14**
47.	Rye	*Secale cereale*	**14**
48.	Barley	*Hordeum vulgare*	**14**
49.	Arabidopsis	*Arabidopsis thaliana*	**10**
50.	Haplopappus	*Haplopappus gracilis*	**4**

CHAPTER – II
CROP SCIENCES

Soil Science

High P uptake efficiency may depress plant uptake of Zn, and therefore cause a reduction in the concentration (density) of Zn in grains of wheats grown in low P (and possibly low Zn) soils.

Mycorrhizae can be divided into two main categories: ectomycorrhizae and endomycorrhizae, of which vesicular arbuscular mycorrhizae are the most widespread in the plant kingdom (Smith and Read, 1997). The mycorrhizal symbiosis is founded on the mutualistic exchange of C from the plant in return for P and other mineral nutrients from the fungus. Influx of P in roots colonized by mycorrhizal fungi can be 3 to 5 times higher than in non-mycorrhizal roots.

Forms of Phosphorus available for Plant Uptake

- The orthophosphates, $H_2PO_4^-$ and HPO_4^{2-}, are the primary forms of phosphorus taken up by plants.
- When the soil pH is less than 7.0, $H_2PO_4^-$ is the predominate form in the soil.

Functions of Phosphorus in Plants

Phosphorus is involved in many plant processes, including:

- Energy transfer reactions
- Development of reproductive structures
- Crop maturity
- Root growth
- Protein synthesis

Soil and Crops

Alluvial soil is a fertile soil rich in potassium. It is highly suitable for agriculture, especially for crops such as paddy, sugarcane and plantain. Red soil has high iron content and is fit for crops like red gram, Bengal gram, green gram, groundnut and castor seed. Black soil is rich in calcium, potassium and magnesium but has poor nitrogen content. Crops like cotton, tobacco, chilly, oil seeds, jowar, ragi and maize grow well in it. Sandy soil is low in nutrient content but is useful for growing trees such as coconut, cashew and casuarinas in areas with high rainfall.

Interveinal chlorosis, a symptom of iron, zinc and manganese deficiencies, evident in yellow parts of leaf. Many micronutrients are used by plants to process other nutrients or work together with other nutrients, so a deficiency of one may look like another (for instance, molybdenum is required by legumes to complete the nitrogen fixation process).

Micro nutrients and Crops

S.No	Micronutrient	Crop	Application (kg/ha)
1.	Zinc sulphate	Rice, Sorghum, Maize, Wheat, Finger millet	25 kg - Soil
		Black gram	2% foliar
2.	Copper sulphate	Sun flower, Ground nut	10 kg - Soil
3.	Ferrous sulphate	Rice, Sorghum, Black gram	50 kg - Soil
		Sugarcane	100 kg - Soil
4.	Manganese sulphate	Soybean, Green gram	1% - foliar
		Finger millet	20 kg – Soil
5.	Borax	Ground nut	10 kg- Soil
		Sunflower	0.2%- foliar
6.	Sodium molybdate	Black gram, ground nut	1% - foliar

Bio fertilizers

- It fixes atmospheric N (20 to 200 kg per ha)
- It solubilize soil P (30 to 50 kg per ha)
- It stimulate plant growth through synthesis of growth promoting substance (Auxin, Vitamin)
- It is a source of renewable energy (bio energy)
- Cost effective compared to chemical fertilizers
- It can be easily produced and applied
- It increase crop yield by 10 to 50 %
- If inorganic form of N is more in the soil than the N fixation in the soil will be less

<table>
<tr><td rowspan="5">N</td><td rowspan="2">Symbiotic</td><td>Rhizobium - Legume</td><td colspan="2">Root - bacteria symbiosis (Pulses)</td></tr>
<tr><td>Anabaena - Azollae</td><td colspan="2">Water fern (Rice)</td></tr>
<tr><td>Associative symbiotic</td><td>Azospirillum</td><td colspan="2">Rice, Ragi, Sugarcane, Maize</td></tr>
<tr><td rowspan="2">Non symbiotic</td><td>Heterotroph</td><td colspan="2">Azotobacter (free living)
For Rice, Vegetables, Commercial crops</td></tr>
<tr><td>Autotroph</td><td colspan="2">BGA - for rice only</td></tr>
<tr><td rowspan="4">P</td><td rowspan="2">P solubilizers</td><td>Fungi</td><td>Aspergillus, Pencillium</td><td rowspan="4">For all crops</td></tr>
<tr><td>Bacteria</td><td>Bacillus, Pseudomonas</td></tr>
<tr><td rowspan="2">P mobilizer or absorber (root fungi symbiosis)</td><td>Ectomycorrhiza</td><td>--</td></tr>
<tr><td>Endomycorrhiza</td><td>VAM</td></tr>
</table>

- N biofertilizer : Azospirillum, Azotobacter, Rhizobium
- P biofertilizer : Phophobacterium

- Plant growth promoting biofertilizer : BGA
- Symbiotic : Rhizobium with legume, Azolla (water fern), Mycorrhiza
- Associative symbiotic : Azospirillum
- Non symbiotic
  a. Free living BGA (Nostac, Anabaena)
  b. Aerobic bacteria (Bacillus, Azotobacter)
  c. Anaerobic bacteria (Clostridium, Rhodospirillum)
  d. Phosphate solubilizing organisms (Phosphobacteria)

Rhizobium

- Associated with legume roots (Pulse, Oil seeds, Tree legumes)
- Isolated by Beijerinck
- Cross inoculation concept given by Fred Baldwin
- Soil should have optimum P, Ca, Mo for optimum N fixation
- Min of 108 viable cells per gram of carrier material (culture) should present for N fixation
- It fixes 25 kg N per ha
- **Nitrogin** : The bio fertilizer preparation with Rhizobium culture

Cultures

Rhizobium leguminosarum	Pea
R. phaseoli	Bean
R. japonicum	Soyabean (slow growing bacteria than other species)
R. trifoli	Clover
R. meliloti	Alfalfa (fix maxi of 300 kg N per half per year)

Azolla

- Reddish colour due to accumulation of Anthocyanin pigment
- Bilobed leaves and long suspended roots are present
- Fleshy upper leaves have cavities which are aerial, chlorophyllous and contain BGA *Anabaena azollae*
- Common cultures in India : Azolla pinnata, A. mexicana & A. nilotica

Adaptation

- Temperature of 20 to 28 C
- Soil should have optimum P
- Standing water of 7.5 cm for 15 days is essential
- Suitable for paddy field
- Fixes 40 to 80 kg N per ha
- Azolla used as green manure (it produce 15 ton GM)

Constrains

- Susceptible to parasites and temperature variation
- Pre flooding is necessary
- P requirement is higher than rice crop

Mycorrhiza

- It is called as fungal roots (Casuarina)
- Ecto mycorrhiza : Hypha are inter cellular (host - Basidiomycetes)
- Ento mycorrhiza : Hypha are intra cellular (host - Orchids, Vanilla, Casuarina)
- Disintegration of hypha within the cell is called Tolypophage or Ptyophage
- Mycorrhiza absorb nutrient from soil and Thiamine + CHO from plants

- It absorbs NPK from soil and supply to plants
- It produces IAA, metabolites and supply it to plant roots
- VAM (Vasicular Arbuscular Mycorrhiza) : they increase surface area of plant roots (VAM is not cultures in pure culture)
- Endotropic mycorrhiza fungi penetrate epidermis and cortex cells of plants roots

Azospirillum

- Azospirillum lipoferum & A. brasilense for Cereals and Grasses
- It fixes 20 to 40 kg N
- It live in close proximity
- Yield increased by 15 to 30 %

BGA (Blue Green Algae)

- It is known as Cyanobacteria (contains chlorophyll)
- It is unicellular, colonial and filamentous type organism
- They are photosynthetic (so need bright sun light)
- It fixes N by Heterocyst
- Suitable for paddy field (effective only in submerged rice)
- It needs temperature of 30 to 35 C, high soil P & low soil N for adaptation
- It synthesis IAA, Vitamins
- It supply oxygen to the plant
- It can reclaim alkali soil
- Apply 10 kg per ha of dried algal flakes
- BGA supplements 25 kg N per

Azotobacter

- Azotobacter chrococum & A. paspali

- It demands high organic matter for N fixation
- It produce IAA, GA, vitamin, nicotinic acid
- It fixes 20 - 40 kg N per ha
- Azotobactor fix N in free soil
- Suitable for paddy, millets, cotton, sugarcane, vegetables, sunflower

Phosphate solubilizing organisms

- Fungi (Aspergillus, Pencillium)
- Bacteria (Phosphobacteria, Bacillus, Pseudomonas, Streptomyces)
- It solubilize the insoluble P into soluble form by secreting organic acid like formic, acetic, lactic, fumaric and succinic acids
- These acids lower the soil pH and bring dissolution of P
- They solubilize 20 to 30 % soil P
- It increase yield by 10 to 20 %
- Suitable for all crops
- It fix no Nitrogen

Genesis of Indian Agriculture Development

1875	IMD at Pune, Maharastra
1904	First Cooperative Credit Societies Act
1905	Imperial Agricultural Research Institute, Pusa, Bihar (now IARI, New Delhi)
1927	All India Radio, New Delhi (1930 in Chennai)
1928	Royal Commission
1929	Imperial Council of Agricultural. Research (now ICAR 1948)
1943	Bengal Famine (caused by *Helminthosporium oryza*)
1945	Grow More Food Programme

1947	Central Tobacco Research Institute, Rajahmundry (AP)
1947	Firka Development Scheme of TN
1950	Sugarcane Breeding Institute, Coimbatore
1951	Bhoodan Movement
1951	All India Rural Credit Survey (1969 All India Rural Credit Review Committee)
1951	Cooperative Movement
1951	First Agricultural Labour Enquiry Committee (1956 Second Committee)
1952	Community Development Programme (Oct 2, Gandhi birth, now Rural Development Programme)
1956	Balwant Roy Mehta Committee - Panchayt raj 3 tier system (1977 Ashok Mehta 2 tier system)
1957	State Seed Farm Project
1957	Central Warehousing Corporation
1957	Fertilizer Control Order
1958	Panchayat Raj System (Rajasthan)
1960	Intensive Agricultural District Programme - (7 district in the country, Thanjavor in TN)
1960	First Agricultural University at Pant Nagar
1963	National Seed Corporation
1964	Intensive Agricultural Area Programme
1965	Indian National Agricultural Policy or Pre Green Revolution period
1965	Agricultural Prices Commission (now 1985 Commission for Agricultural Costs and Prices)
1965	National Demonstration Scheme (to be conducted by scientists in farmers field)
1965	Food Corporation of India
1966	Seed act
1966-67	Green Revolution & HYV Programme

1968	Seed Rules
1968	Insecticide Act
1970	National Commission on Agriculture
1970	Small Farmers Development Agency & Marginal Farmers and Agricultural Labours
1971	Drought Prone Area Programme (DPAP)
1971	Agro Service Centre Scheme
1973	Department of Agricultural Research and Education (DARE)
1974	T & V system (TN was 11th state to implement this system during 1980)
1974	Command Area Development Programme
1974	First KVK at Pondicherry under TNAU (1977 - other KVKs)
1975	Regional Rural Bank (Oct 2 - Gandhi Jeyanthi)
1975	Operational Research Project
1978	Integrated Rural Development Programme (IRDP)
1979	TRYSEM
1979	Lab to Land Programme (ICAR Golden Jubilee Year)
1982	NABARD
1984	Cooperative Societies (1904 Coop Banks 1904)
1985	Comprehensive Crop Insurance Scheme (1999 National Agricultural Insurance Scheme)
1986	NWDPRA, Integrated Watershed Development Programme
1986	Revised 20 Point Programme (Announced by Indra Gandhi)
1986	TANWA Project (aided by DANIDA)
1989	Jawahar Rojgar Yojana
1989	Service Area Approach (adopted by Commercial Banks)

1991	Economic Reforms, Liberalization, New Economic Policy, New Trade Policy
1991	Tamil Nadu Agricultural Development Project (TNADP)
1993	Prime Minister's Rojgar Yojana
1995	Globalization & WTO
1995	Agreement on Agriculture (AoA)
1998	National Agricultural Technology Project (NATP)
2000	New Agricultural Policy (National Policy on Agriculture)
2000	Technology Mission on Cotton
2001	Technology Mission on Coconut
2001	Protection of Plant variety & Farmers Right Act
2002	Multi State Cooperative Society Act
2002	National Seed Policy
2004	International year of Rice (Announced by FAO - slogan is *rice is life*)
2004	National Commission for Farmers

CHAPTER – III
PLANT PROTECTION

Pesticides or Entomo pathogens: Disease causing organisms on insect pest

A. Bacteria

- *Bacillus thurungiensis* (Bt) » cotton boll warm, rice stem borer, fruit borer of tomato, brinjal, bhendi
- Pseudomonas » root rot disease
- Bt: It has been isolated from *Heliothes armigera* from Mysore (1955) and locally produced by *Solid State Fermentation Technology*. It's commercial powder or liquid form applied @ 0.3 - 0.5 kg per ha.

B. Fungus

- *Trichoderma viridi* » on root rot, wilt diseases
- Verticillium » on scale, aphids, root rot, wilt disease
- *Beauveria bassiana* » borers of sugarcane & cotton, rice hispa
- *Metarhizium anisopliae* » rice stem borer, pyrilla, spodoptera
- *Fusarium solani* » cotton boll warm
- Muscardine » on root rot, wilt diseases
- Cephalosporium
- Entomophthora
- Protozoa (Nosema)

C. Virus

- Granulose Virus Chilo » sugarcane top borer
- GV Achaea janata » castor semi looper (Aa NPV also used for bio control)
- Ha NPV » *Heliothis armigera*
- Sl NPV » *Spodoptera litura*

- *Baculovirus oryctus* » Rhinoceros beetle of coconut
  a. Bt gene incorporation in : cotton, rice, maize, tobacco, tomato, potato
  b. Apple wooly aphids are controlled by *Aphelinus mali* (parasite)
  c. *Bacillus thuringiensis* (bacteria) controls most of the Lepidopteran pest
  d. NPV for spodoptera litura, helicoverpa armigera (heliothis armigera)
  e. Cytoplasmic polyhedrosis virus and Granuloses virus (parasites)
  f. *Metarrhizium anisopliae* (fungi) for sugarcane pyrilla
  g. Entomophthora for butterflies and leaf hoppers
  h. *Beauveria bassiana* for castor semi looper
  i. Cephalosporium, Aspergillus also used for bio control

National Centre for IPM - New Delhi

Project Directorate of Bio control - Bangalore

Vectors

Most of the viruses are transmitted by *Hemipteran* insects (because it has most suitable mouth parts)

Hemiptera: Aphids, leaf hopper, plant hopper, white fly, mealy bug, Other vectors are thrips (Thysanoptera) and some grass hopper (Orthoptera) and beetles

- Biting and chewing mouth parts : beetles, caterpillars
- Piercing and sucking mouth parts : white fly, aphids, leaf hoppers

Table 1 Pest and Diseases of Plants

Crop	Disease	Casual organism	Vector
Rice	RTV	Virus	GLH or Green Jassid *(Nephotettix virescense)*
	Grassy stunt	Mycoplasma	BPH *(Nilaparvada lugans)* BPH produce Hopper burn
	Yellow dwarf	Mycoplasma	Green Leaf Hopper
Banana	Bunchy top	Virus	Aphids *(Pentalonia nigronervosa)*
Bhendi	Yellow Vein Mosaic	Virus	White fly *(Bemisia tabaci)*
Blackgram Greengram Red gram Lab lab	YMV	Virus	White fly *(Bemisia tabaci*)
Black gram Green gram	Leaf curl virus	Virus	Thrips
Cowpea	YMV	Virus	Aphids *(Aphis gossypi)*
Red gram	Sterility mosaic	Virus	Eriophid mite *(Aceria cajani)*
Groundnut	Ring mosaic or Bud blight or Bud Necrosis	Ring mosaic virus	Thrips
	Rosette	Virus	Aphid
Sesame	Phyllody (Little leaf)	Mycoplasma	Leaf hopper (Jassid)
Brinjal	Little leaf	Mycoplasma	Plant hopper (Jassid)

Chilli	leaf curl	Tobacco leaf curl virus	Thrips
Sugarcane	Grassy shoot	Mycoplasma	Aphids *(Aphis maidis)*
	Mosaic	Virus	Aphid
Cardamom	Katte disease	Virus	Aphids *(Pentalonia nigronervosa)*
Tomato	Leaf curl disease	Virus	White fly
	Spotted wilt	Virus	Thrips
	Mosaic	Virus	Seed born
	Black ring spot	Virus	Nematodes
	Big bud	Mycoplasma	Leaf hopper
Citrus	Rubbery wood	Mycoplasma	
	Leathery leaf or Leaf curl or Yellow corky vein or Mosaic or Bud union	Virus	
Papaya	leaf curl	Nicotiana virus 10	
Ber	Witches broom	Mycoplasma	Grafting
Coconut	Kerala wilt	Mycoplasma	Soil born

Angiospermic (flowering plants) parasites

Parasites establish relationship with their host through **Haustoria** to derive nutrition from vascular system of host plants

But parasites has chlorophyll

1. **Striga** (Partial root parasite): It attacks roots of sorghum, maize, sugarcane

   Control measures: Flood irrigation, 2,4 D application & raising trap crop like Sudan grass

2. **Orabanche** (Total root parasite) : It attacks roots of Solanaceous & Cruciferous plants
   Ex. Roots of tobacco, brinjal, tomato, cauliflower
   Control measures: Raising of trap crop (Chilli)
   Alkaloids present in this trap crops are reduce the germination of parasitic weeds
3. **Loranthus** (Partial stem parasite) : It attacks fruit, avenue and forest trees
   Control measures: Injection of $CuSO_4$ and 2,4 D
4. **Cuscuta** (Total stem parasite) : It attacks all crop plants except cereals

Bordeaux mixture

Oldest fungicide

Discovered by Millardet 1882 (Botany professor, University of Bordeaux, France)

Found out in downey mildew of grapevine

Constituents: Lime 1 kg + Copper sulphate 1 kg + Water 100 lit ($CuSo_4$ + $CaCo_3$ + water)

Add copper sulphate solution into lime solution and not vice versa

Lime acts as neutralizing agent

Use prepared mixture immediately

If you want to keep it for another 24 hrs add jaggery 1 g per lit of water

BM is phytotoxic to HYV of rice, maize and apple

Burgundy mixture : $CuSo_4$ + Na_2Co_3 + Water (used for Cu sensitive crops)

Bordeaux paste: As fungicide used for controlling stem bleeding of coconut

Elemental sulphur

Used for controlling Powdery mildew, Downey mildew, Rust, Tikka disease, Potato scab

Sulphur fungicide is phytotoxic to cucurbits

Dithane S 31 : Used for controlling cereal rust

Karathane : Used for controlling Powdery mildew and Downey mildew

Chemical name	*Trade name*
Copper oxy chloride	Fytolan, Blue copper
Carboxin	Vitavax
Oxycarboxin	Plantvax

Antibiotics

The are produced by microbes in small quantity

This antibiotics inhibits the growth of other microbes

Most of the antibiotics are developed from **Actinomycetes** *(Streptomyces spp)* and some from fungi

Anti fungal antibiotic : Aureofungin

Ex. Citrus gummosis cuased by fungi *Phytophthora sp*. It if controlled by Aureofungin

Anti bacterial antibiotic: Streptomycin sulphate

Marketed as Agrimycin 100 (Streptomycin 15 % + Terramycin 75 %)

Anti mycoplasma antibiotic: Tetracycline Produced from Streptomyces sp

Plant Quarantine

The consignment (plant materials, seeds) imported should be accompanied by a **Phytosanitory Certificate** (PC)

The following plant materials are restricted for import

- Cotton seed for boll weevil
- Sugarcane sett for sugar weevil
- Coffee seed for berry borer

The following communities should be accompanied by PC when exported

- Pepper, Cardamom, Tamarind

Madras Agricultural Pests and Diseases Act: The act was enacted in 1919 by the Madras Govt.

It was the first state to enact such act in this country

In India, 16 quarantine stations are operating (8 sea port + 6 air port + 2 land frontier) through out the country

In Tamilnadu : Chennai and Tiruchirappally

Exclusive quarantine: There is a complete restriction of movement of the plant material of a particular crop from a specified region

Ex. Banana seed material import is completely restricted from Srilanka

India does not import rubber seed or plant from South America and West Indies

Regulatory quarantine: The plant material are allowed after inspection and certification (Phytosanitory certificate). In India, Potato seed tubers are not allowed for sale from Nilgiri's to prevent the Golden nematode spread *(H. rostochinensis)*

Domestic quarantine is operated for 2 insects : Fluted scale and Sanjose scale

Domestic quarantine for 3 diseases : Wart disease of Potato, Bunchy top of Banana & Mosaic virus of Banana

Table 2 Introduced diseases into India

Diseases	Casual organism (Introduced)	Country from which introduced
Bunchy top of banana	---	Srilanka
Coffee leaf rust	*Hemeilia vastatrix*	Ceylon
Potato Late blight	*Phytophthora infestans*	Europe
Grapes Downey mildew	*Plasmopora viticola*	Europe
Maize Downey mildew	*Peronoselero spora*	Java

Potato Golden nematode	---	Europe
Potato Wart disease	---	Netherlands
Paddy Blast	*Pyricularia oryza*	South East Asia
Rubber Powdery mildew	*Oidium haveae*	Malaysia

Entomo Pathogenic Fungi (EPF)

EPF are a major component of integrated pest management techniques as biological control agents against insect pests and other arthropods and are an integral part of myco-insecticides in horticulture, forestry and agriculture. Some of the commonly used EPF are as follows.

- ***Beauveria bassiana*** - These enter the host insects' body through food or in contact with the host cuticle and reproduce inside the insect body. It produces toxins namely beauvericin, bassianocide etc. inside the host body causes paralysis of the host insects and ultimately kills the insects within four or five days. They are used particularly to control sucking pests and caterpillars infesting crop plants. These fungi are used to control the caterpillars of yellow stem borer and leaf folder of rice, white grub of groundnut, sugarcane pyrilla, coconut rhinoceros beetle, caterpillars of pulses, tomato and cotton, diamond back moth, leaf eating caterpillars of tobacco and sunflower etc.
- ***Lecanicelium lecanii*** - This beneficial fungus is mainly used to control whiteflies, aphids, thrips, brown plant hopper, scale insects, mealy bugs and other sucking insect pests of crop plants.

- ***Metarrhizium anisopliae*** - This pathogenic fungus is used to control mainly coconut rhinoceros beetle, groundnut cut worm, rice brown plant hopper, diamond back moth and early shoot borer, top shoot borer and internode borer of sugarcane.
- ***Nomuraea rileyi*** - It is used to control pod borers, cut worms, cabbage borers etc.
- ***Hirsutella thompsonii*** - These fungi are used to control different hoppers and bug pests, whiteflies, red mites etc.
- ***Paecelomyces fumosoroseus*** - This fungus is used to control yellow and red mite, whiteflies etc.

CHAPTER - IV
FOOD AND NUTRITION

Seed proteins

Seed proteins have been divided into four fractions **based on their solubility** in different solvents.

1. Water soluble protein : Albumins
2. Salt soluble protein : Globulins
3. Alcohol soluble protein : Prolamins
4. Alkali or acid soluble protein : Glutelins

Prolamins and glutelins constitute the bulk of the protein of most cereals, while in legumes about 80 per cent of the total protein consists of globulins. So, sambar traditional Indian vegetarian food/dishes can be slightly salty for enhancing the availability of legume protein into our body.

Oat (*Avena sativa* L.) protein is composed mainly of globulin (80 per cent) with prolamin (avenin) representing only 12 to 20 per cent of the total protein. Consequently Oat protein is in general nutritionally superior to other proteins except rice.

Fortification

Adding vitamins artificially to the foods that are eaten. This process termed fortification. Eg: Milling with vitamins such as Thiamine, riboflauin, miacin and pyridoxine to wheat flours.

Addition of Vitamin D to milk; Addition of Iodine with table salt

It under nutrition is seen, the initial stage of severe under nutrition is called marasmus. It under nutrition continues, the child may progress to the more serious disease Kwashiorkor

(West African word meaning "the disease the child gets when another baby is born".

Malnutrition may arise in one of four ways

- There is simply not enough food available and the consequence is under nutrition (Marasmus and kwashiorkor) - Protein energy - mal nutrition (PEM).
- The diet may lack one essential nutrient or more, giving rise to deficiency diseases such as Xerophthalmia (due to deficiency of Vitamin A), beriberi (Vitamin B1), Pellagra (Nicotinic acid Vitamin B4), Scurvy (Vitamin C), rickets and Osteomalaria (Vitamin D), goiter (Iodine) and several kinds of anaemia (iron, folic acid).
- A person may have a condition are an illness, either genetic or environmental (eg. Infections) in origin, that prevents him from digesting food properly or from absorbing some of its constituents, that is called secondary mal nutrition.
- Too many calories or an excess of one or more components of a reasonable diet are consumed; this condition is oven nutrition.

Severity of Vitamin A deficiency

Night blindness ⟶ Xerophthalmia ⟶ Keratomalacia

(First symptom) (Infant child) (Severe symptom)

(Adult)

Some of the β - carotene provided by food is converted into Vitamin A and therefore it is called Provitamin A. The total Vitamin A value, obtained from both Vitamin A (retinol) itself and the converted carotene, is referred to as Vitamin A activity.

- Kwashiorkar is an African world from Ghana which literally means "rejected child". It develops when babies are weaned from protein - rich breast milk to other diets which are deficient in protein and energy.
- Pellagra (Caused due to deficiency of Vitamin B4 = (Niacin)
- Pellagra is common amongst people living on a maize diet, because maize interferes with the absorption of niacin in the body.
- Calcium and (Magnesium) deficiency is common amongst people living on a pearl millet or cumbu diet, because phytic acid present in pearl millet interferes the absorption of calcium in the diet.

Vitamins are classified as either **fat soluble** (vitamins A, D, E and K) or **water soluble** (vitamins B and C). This difference between the two groups is very important. It determines how each vitamin acts within the body.

Fat soluble vitamins: Vitamins A, D, E and K
Water soluble vitamins: Vitamins B and C

The fat soluble vitamins are soluble in lipids (fats). These vitamins are usually absorbed in fat globules (called chylomicrons) that travel through the lymphatic system of the small intestines and into the general blood circulation within the body. These fat soluble vitamins, especially vitamins A and E, are then stored in body tissues. Although, when it comes to vitamin pills, few of them are water soluble that bypass the body system without harming much but, the oil-soluble ones retain in the body fats and are harmful.

Vitamin A Inhibits K

Your body needs vitamin A to keep your eyes healthy, regulate gene expression, produce hormones, make red blood cells and support a healthy immune system. The Institute of Medicine recommends adult men consume 900 micrograms and adult women get 700 micrograms of vitamin A per day. If taken in large doses, excess vitamin A can inhibit your body's absorption of vitamin K, a vitamin essential for normal blood clotting. Although this is the only recognized vitamin-to-vitamin interaction that affects absorption, vitamins can also affect the absorption of minerals.

Vitamin E Inhibits K

Vitamin E, a vitamin that protects your cells from damage and supports a healthy immune system, does not inhibit the absorption of vitamin K, but it does inhibit its activity. Studies suggest that vitamin E can prevent the formation of blood clots and reduce your risk for heart attack or venous thromboembolism, according to the Office of Dietary Supplements. Vitamin E inhibits blood clot formation by inhibiting the actions of enzymes activated by vitamin K. In healthy individuals, large doses of vitamin E -- doses in excess of the tolerable upper limit set as 1000 milligrams per day -- therefore increase your risk of uncontrolled bleeding.

Zinc Deficiency inhibits Vitamin A

Zinc, an essential mineral found in meats, seafood, nuts, beans and grains, activates approximately 100 enzymes in the body and plays a role in the production of proteins, wound healing, cell division and DNA synthesis and supports a healthy immune system. A zinc deficiency -- getting less than the recommended 11 milligrams daily for men and 8 milligrams for women over a period of time -- inhibits vitamin A activity. Your body stores excess vitamin A, which is a fat-soluble vitamin, in the fat cells and liver until needed. Enzymes activated by zinc release vitamin A from storage. A zinc deficiency also decreases the amount of the protein known as retinol binding protein that transports vitamin A to the tissues throughout the body

Casimir Funk, who coined the name Vitamine, denoting an amino essential for life. (amino from rice husk and polishing that alleviated the symptoms of the disease beriberi)

Folic acid (Latin folium, "Leaf"), first found in spinach leaves.

Vitamin E (tocopherol; Greek tokos, "child birth").

Vitamin K (K for Danish, Koagulation)

The compound dicumarol, an analog of vitamin K. **Dicumarol** is used in clinical medicine to prevent clotting in blood vessels.

Table 1 Vitamins and their coenzyme forms

S. No.	Type	Coenzyme or active form
A.	**Water - soluble:**	
1.	Thiamin	Thiamin pysophosphate (TTP)
2.	Riboflavin	Flavin mono nucleotide (FMV) Flavin adenine dinuceleotide (FAD)
3.	Nicotinic acid	Nicotinamide adenine dinucleotide (NAD) Nicotinamide adenine dinucleotide phosphate (NADP)
4.	Pantothenic acid	Coenzyme A (CoA)
5.	Pyridoxine	Pyridoxyl phosphate
6.	Biotin	Biocytin
7.	Folic acid	Tetrahydrofolic acid
8.	Vitamin B_{12}	Coenzyme B_{12}
9.	Lipoic acid	Lipoyllysine
10.	Ascerbic acid	--
B.	**Fat soluble**	
1.	Vitamin A	11-cis-retinal
2.	Vitamin D	1, 2, 5-Dihydroxycholecalciferol
3.	Vitamin E	--
4.	Vitamin K	--

Food preservation

1. Meat, Fish and their products are dehydrated by a process called smoking. Small pieces of meat are first rubbed with salt and then hung from the ceiling or kept on wooden rafters. These pieces are then dehydrated with the heat and smoke of the fire lit below.
2. Chemical preservation
   Two common chemical preservatives permitted for use in our country are, Benzoic acid (also benzoates) and Sulphur dioxide (including sulphates)

Benzoic acid is an effective agent for preservation. Its sodium salt, sodium benzoate, is used for preservation of fruits, juices and squash because it is soluble in water. Sodium benzoate of 0.06% to 0.1% concentration is quite adequate for preserving fruit juices.

Potassium meta-bisulphate is also an effective preservative when added to fruit juices or squash, the potassium radical reacts with the acid of the juice, forming the corresponding potassium salt and sulphur dioxide. It is used for the preservation of colorless fruits like apples, litchis, raw mango chutney, and lemon squash. Sodium bicarbonate (baking soda) destroys Vitamins B and C.

Food poisoning

- The bacteria *Salmonella* which causes food poisoning, is derived from farm animals through contaminated meat, egg. Milk and milk products. Multiplication of this bacteria in our intestines causes nausea vomiting, diarrhea, and pain in the abdomens.
- The bacteria *Staphylococci* produce a toxin, which is heat resistant.

- A rare but dangerous form of food poisoning, called Botulism is caused by the bacteria *Claustridium botulinum.*
- **Aflotoxin** is a toxin produced during the growth of the fungus. Aspergillus on grains (wheat, rice, maize and groundnut) and vegetables.

CHAPTER - V
ANIMAL SCIENCES

The most important exotic dairy cattle breeds in India are Jersey and Holstein Friesian (HF).

- **Poultry Farms:** Farms having more than 1000 birds will be considered as Poultry Farms. Farms having **only** Emu/Ostrich birds will be considered as farm if having more than 20 birds.
- **Livestock:** Cattle, Buffalo, Mithun, Yak, Sheep, Goat, Pig, Horse, Pony, Mule, Donkey Camel, Pig, Dog, Rabbit and Elephant.
- **Poultry:** Fowl, Duck, Emu, Turkeys, Quail and other poultry birds.
- **Sheep:** The important exotic breeds are Merino, Rambullet and Carridale. Sheep have been classified broadly in two age groups namely; under 1 year and 1 year & above.
- **Goat:** goat is said to be 'poor man's cow' & provides a dependable source of income to 40% of rural population below the poverty line in India. Goat provides milk, meat, hides, manure & pashmina. Goat have been classified broadly in two age groups namely; 'under 1 year' and '1 year & above' for both male and female.

Goat differs from sheep in having 60 chromosomes in comparison to 54 in sheep.

Pashmina refers to a type of fine Kashmir wool and the textiles made from it

- The milch type goats are more in temperate zone. Dual type / meat type goats are mostly located in subtropical & tropical Asian & African countries. The nutritive value of goat milk does not differ from that of cow milk. The calcium, phosphorus & chlorine content are higher than cow milk. The meat of goat is called chevon, it is low fat content. Dressing % is 43-53% & optimum slaughter age is 6-12 months for best quality meat.

- **Chegu & Chengthani** breed produce pashmina. Mohair is obtained from some of goat breeds. Goat manure has more nutrient value than cow's manure. Goats are polyesterous animal. The gestation period is 145-153 days & 18-21 days oestrus cycle. The duration of oestrus is 24-48 hours. The average life span under ordinary condition is 12 years in goat in India.

Exotic breeds of goat

- Toggenburg = Toggenburg valley of Switzerland = 800-1000 liters of milk per lactation of 150-200 days.
- Saanen = Saanen valley in Switzerland = 1000 liters milk per lactation.
- French Alpine = alphas place of France = 800-1000 liters milk per lactation.
- Nubian = Nubian place of South-east Africa & South Asia = 1200 liters milk per lactation.
- **Angora** = Asia Minor.

Table 1 Indian Breeds of Goat

S. No.	Name of Traditional breed	Region	Economic Importance
1.	Assam hill goat	village of Khasi, Naga, Lushal hills	Meat
2.	Barbari	Agra, Aligarh, Etah, Delhi, Haryana	Dual purpose (milk & meat)

3.	Black Bengal	village of Orissa & Bengal	Meat
4.	Chegu	Spiti, Kashmir, Tibetian mountians	**Pashmina** (textile wool)
5.	Kutchi	Cutch, Kathiawad	Meat
6.	Gaddi	Hills of Himachal pradesh	Hair
7.	Ganjan	Orissa & parts of west Bengal	Meat
8.	Jamunapuri	Chambal rivers, Chakarmagar area of U.P	Dual purpose (milk & meat)
9.	Chang thangi	Hilly tract of Leh & Ladakh	**Pashmina** (textile wool)
10.	Bakharwal	Hilly tract of Kashmir	Hair
11.	Malbari & Tellicheri	Kusum, Kottayam, Calicutt, Ponnani	**Milk**
12.	Ranwari	Jodhpur, Mehsana & Marwar	Dual purpose (milk & meat)
13.	Osmanabadi	Andhra pradesh	Dual purpose (milk & meat)
14.	Sirohi	Sirohi & Udaipur	Dual purpose (milk & meat)
15.	Mehsana	Baroda & Mehsana	**Milk**
16.	Surti	Surat (Gujarat)	**Milk**
17.	Zalwadi	Kathiawad, western Mehsana	Dual purpose (milk & meat)
18.	Zakhrana & Prabatsar	Ajmer & Nagore	**Milk**

Pig: The important exotic breeds are Landrays and Yorkshire. Pigs have been broadly grouped in two age groups namely; 'under 6 months' and '6 months & above'. Horses and Ponies look alike. Those having height of 4 feet 8 inches and above are categorized as horses and the rest are the Ponies. A **mule** is the offspring of a male donkey and a female horse.

Backyard poultry: If a hen/duck lays more than 100 eggs in a year then it is categorized as 'improved' hen/duck or else it is categorized as 'desi'. Broilers may be covered under 'improved' variety

CHAPTER - VI
GENERAL SCIENCES

BGR - Blue, Green and Red colors (primary colors)

Eg: Red + blue	=	Magenta
Blue + green	=	Cyan or peacock blue
Red + green	=	Yellow

- John Dalton, founder of the atomic theory was colour blind and could not distinguish even between primary colours. As a result, this condition of colour blindness was known in the medical profession for quite some time as *Daltonism*.
- Animals differ in their colour perception, the bee has retinal cone cells that enable it to see colours beyond indigo and violet, namely ultra violet. We human being are ultraviolet blind.
- The chicken eye retina has mostly cones and very few rod-shaped cells. The chicken therefore needs bright light to see, that's why it wakes up with the sun and roosts by sundown.
- The famous Sahara desert was formed as a result of overgrazing by progressively increasing tribe of herbivores. This happened because of capture of lions by Romans which reduced the predator population. Similarly, overgrazing also resulted in making most of Rajasthan a desert.
- **Caprification** refers to an artificial method of pollinating certain varieties of cultivated figs.
- Folic acid first found in Spinach (*Spinacia oleracea*) leaves.
- Potato plant was introduced from the new world to old world.

- **Cauliflory:** production of flowers on old stem from new buds.
- The rate of Plants' transpiration during storms is nil.
- In grasses, stem increases in length by intercalary meristem
- The first plant in which chromosomal basis of sex determination was discovered is Melandrium.
- A chemical which causes rootings in cuttings - IBA
- CO_2 acceptor in C_3 plants is RUDP (Ribulose 1, 5 diphosphate)
- CO_2 acceptor in C_4 plants is PEP (Phospho enol pyruvate)
- FAD or FMV is a coenzyme, which vitamin is incorporated into its structure? Vitamin B_2 (Riboflavin)
- One of the plants introduced from the new world to old world is Potato.
- Annual rings are distinct in plants growing in temperate regions.
- Poly embryony : Citrus (type of apomixis)
- Cauliflory: Production of flowers on old stem from new buds.
- Triple antigen is a vaccine against Tetanus, Whooping cough and diphtheria.
- Potometer : Transpiration
- The hormone which controls the outflow of potassium from guard cell is abscisic acid (ABA).
- Which of the following tissue is widely used for culturing animal viruses - Chorio allontoic membrane?
- The smallest virus is foot and mouth virus of cattle (Virus size: 2nm and 300 nm)
- Viral - DNA attached to bacterial DNA is called Provirus.

- Which part of the plant cannot be infected by virus? Shoot apex (meristem)
- Antibiotics fail to act on viruses because they have no metabolism of their own.
- A virus infecting three different host plant is LPP-1.
- The tonoplast is differentially permeable.
- During photosynthesis oxygen in glucose comes from carbon dioxide.
- Color indicator method to find out transpiration is by using filter paper dipped in cobalt chloride solution. What is the color when it is in dry (Blue).
- What is the rate of transpiration during storms? Nil
- When a branch of a tree is cut, the wound gets healed by the formation of phellogen producing cork.
- In grasses stem increases in length by intercalary meristem.

Bacteria

- First observed by Van Leeuvenhock (1683)
- Named by Ehrenberg (1829)
- Movements: by locomotion flagella
- Monotrichous: Single flagellum at one end
- Lophotrichous: Many flagella at one end.
- Amphitrichous: flagella at both the ends.
- Peritrichous : Flagella along the entire length of the cell.
- Atrichous: without flagella

Morphological types

Based on movements and nature of cell wall.

- Eubacteria: move with the help of flagella and possesses a thick rigid cell wall.

- Myxobacteria: Moves by gliding
- Spirochadetes: Moves by axial filaments - thin flexible cell wall.

Shapes: (Eubacteria)

Coccus: Spherical
Bacillus: rod shaped
Uibrio: Curved rods
Spirillum: Spiral

Actinomycetes are mycelia Eubacteria: Antibiotics production.

- Some of the bacteria produce a substance which emits light on oxidation. The substance which is responsible for luminescence is luciferin.
- Chloromphenicol and erythromycin are broad spectrum antibiotics produced by streptomyces.
- Streptomycin is produced by *streptomyces grisens*.
- Water borne diseases are Typhoid, cholera and dysentery.
- Non leguminous N_2 fixing bacteria is *Azotobacter paspali*.
- *E.coli* cell wall component is muramic acid.
- Some strains of *E.coli* with a particular type of extra chromosomal DNA (factor) have been found to have a greater tendency to mate with the strains that lack it. This factor has been called as F-factor.
- Penicillin inhibits cell wall synthesis, thereby inhibiting bacterial multiplication.
- An extra-chromosomal DNA fragments has been found in several bacteria which replicates itself independently or gets integrated into the bacterial chromosome. This is called episome.

- Bacteria can not survive in a highly salted pickle because - they get plasmolised and are consequently killed.
- The bacteria which reduce the fertility of soil is *Bacillus denitrificans*.
- The bacterial photosynthesis is does not involve oxygen evolution.
- Bacteria are considered primitive to fungi because they have naked DNA.
- Retting is caused by bacteria due to their property of dissolving living contents of cells.
- Botulism is a fatal human food poisoning caused by a saprophytic bacteria Clostridium, which is not a fresh water form:
- *Bacillus subtilis* are hay bacteria.
- Rickettsial is a group of bacteria.
- Typhus is caused by rickettisal.
- Syphilis is the disease caused by Spirochactes.
- Gram positive bacteria is less resistant or more susceptible to antibiotics.
- The red pigment present in the root nodules of leguminous plants is known as leghaemoglobin.

Algae

- Agar - agar is produced from red algae (Red sea)
- Agar - agar is obtained from Gelidium.
- What is the most common method of reproduction in blue green algae: Akinites.
- In Eastern Asia one of the red algae is cultivated on the coast and is eaten under the name "Nori". This is obtained from Porphyra.

- Euglena is often included in the animal kingdom because it has no cell wall.
- Which one of the algae store mannitol, a sugar alcohol, as reserve food material? Brown algae.
- Which of the following algae store its photosynthetic product as oil and not as starch: Xantholhyceae.
- A multi nucleate, aseptate plant body is called coenocytes.
- Reproduction with the help of hormogonia is found in cyanobacteria.
- The laruicidal algae: Oscillatoria.
- Pteridophyta: the first vascular plants.
- Gymnosperms: lycas is dioeious, i.e. male and female cones borne on two separate plants.
- Gymnosperms lack in true xylem vessels.
- Canada balsam is an oleoresin obtained from *Abies balsamea.*
- Turpentine is obtained from *Pinus australis*.
- Cedar wood oil is obtained from heartwood of Juniperus.
- The endosperm in Gymnosperm is haploid.

Angiosperms:

Insectivorous plants		Scientific name
1. Sundew plant	:	Drosera
2. Pitcher plant	:	Napenthes
3. Venus' fly trap	:	Dionea
4. Bladder wort	:	Utricularia

Seed germination

Soaking seed in water triggers germination. Embryo releases Gibberelic acid (GA) which reaches the aleurone cells, stimulating them to synthesize hydrolytic enzymes *amylases* for the digestion of starch that digest nutrients stored in endosperm the resulting sugars, amino acids and nucleotides become available for the growth of the embryo.

Tetrazolium test (TZ)

Tetrazolium test is a useful test for rapidly obtaining an indication of living cells. The living cells are made visible by reduction of an indication dye. Tetrazolium salt (2,3,5 triphenyl tetrazolium chloride) + living cells gives triphenyl formation (red, stable and non diffusible substance).The living cells accept hydrogen from tetrazolium salt. Then the colorless indicator tends to become red stable colour.

Amino acids

The first amino acid isolated from a protein hydrolyzate was glycine, obtained in 1820 from gelatin by H. Braconnot.

A very widely used reaction of the amino groups, the ninhydrin reaction, can be utilized to estimate amino acids quantitatively in very small amounts. On heating, amino acids reacts with two molecules of ninhydrin to yield an intensely colored product. A purple color is given in the ninhydrin reaction by all amino acids and peptides having a free --- amino groups, whereas proline and hydroxyl proline, in which the ---- amino group is substitute yield derivatives with a characteristic yellow color.

Three amino acids - tyrosine, tryphtophan and phenylalanine - absorb light significantly in the ultra violet. Since most proteins contain tyrosine residues, measurement of light absorption at 280mm.

Cystine absorbs weekly at 240mm due to its disulfide group.

(Note : Ultra violet (UV) absorption 220nm) 180-220 nm.

The important reaction of amino group is with cyanate, to yield carbamoyl derivatives. This reaction has been used to modify the properties of Sickle-cell hemoglobin to make it more nearly like normal adult hemoglobin.

The first general theory of chemical catalysis, published in 1835 by J.J. Berzelius named the enzyme as diastase.

An enzyme was first isolated in pure crystalline form. This was accomplished by J.B.Sumner in 1926 for the enzyme urease, isolated from extracts of the jack bean.

International classification of enzymes

1. Oxido - reductases (Oxidaton - reduction reaction)
2. Transferases (transfer of functional groups)
3. Hydrolases (hydrolysis reactions)
4. Lyases (addition to double bonds)
5. Isomerases (Isomerization reaction), and
6. Ligases (formation of bonds with ATP cleavage)

Biological magnification

In 1942, DDT was spread extensively in and around the Lake Michigan in N. America. This was done to kill mosquitoes and consequently eradicate Malaria. By the early sixties, it was found that the number of pelicans (a bird) was going down at an alarming rate. Nobody know why. Then the environmentalists realized that the eater in the lake retained 0.2 ppm of DDT (1 ppm is 1/1,00,000). The plankton (plants and animals) in the lake retained 77 ppm of DDT in their bodies, and the fish that are the plankton retained about 1700 ppm of DDT in their bodies. The effect of such a high level of DDT in the body of the pelicans resulted in very thin shells in their eggs. Hence, even before the egg could hatch the thin shells broke off. This reduced the population of pelicans alarmingly. The above example shows how harmful chemicals enter and get concentrated in a food chain. This process is called Biological magnification.

Effect of Acidic gases on Historical monuments:

1. **Acidic gases**

Sulphur dioxide (SO_2) and nitrogen dioxide (NO_2)

These gases combine with water vapour present in the air and form the respective acids, finally acid rain get.

$2SO_2 + 2H_2O \longrightarrow 2H_2SO_4$ (Sulphuric acid)

$2NO_2 + H_2O \longrightarrow 2HNO_3$ (Nitric acid)

Since the acids so formed affect and damage the marble stone (CaCO3), there is a possibility of the Taj Mahal which is made of marble getting damaged.

$CaCO_3 + H_2SO_4 \longrightarrow CaSO_4 + CO_2 + H_2O$

$CaCO_3 + 2HNO_3 \longrightarrow Ca(NO_3) + CO_2 + H_2O$

Harmful effects of carbon monoxide

When oxygen combines with the hemoglobin of flood, it forms oxyhemoglobin. Similarly, when carbon monoxide combines with the hemoglobin of blood, it forms carboxy hemoglobin (CoHb). The affinity of carbon monoxide for the hemoglobin of blood is 200 times more than that of oxygen. Therefore, even a very small amount of carbon monoxide is sufficient to form CoHb. Carboxy hemoglobin adversely affects the oxygen carrying capacity of hemoglobin, i.e., latter becomes oxygen deficient. The deficiency of oxygen in blood many cause headache, dizziness, cardiac and pulmonary changes and even death.

Alloy is a mixture of metal with other metal or non-metal.

S. No.	Alloy	Constituents
1.	Steel	Iron, Carbon
2.	Stainless steel	Iron, Nickel, Chromium
3.	Brass	Copper, Zinc
4.	Bronze	Copper, Tin
5.	Solder	Lead, Tin
6.	German silver	Copper, Nickel, Zinc
7.	Duralium	Aluminum, Copper, traces of Magnesium and Manganese.

General Science (Questions and Answers)

1. Who is honored as Father of Modern Chemistry?
   Answer: Antoine Lavoisier
2. Who invented periodic table?
   Answer: Dmitri Mendeleev
3. Which is the most abundant gas in the earth's atmosphere?
   Answer: Nitrogen (78.08%)
4. Which gas evolved from paddy fields and marshes?
   Answer: Methane
5. Which toxic element present in automobile exhausts?
   Answer: Lead
6. Which gases cause acid rain?
   Answer: Sulphur dioxide, Nitrogen oxides
7. Which metal is used in the making of microchips?
   Answer: Silicon
8. What is the toxicity caused by silicon called?
   Answer: Silicosis
9. Which polymer is used to manufacture electric switches, computer disks etc.?
   Answer: Bakelite

10. With which polymer, the cabinets of radio and TV made of?
    Answer: Polystyrene
11. Which chemical is used to make rain coats?
    Answer: Poly Vinyl Chloride (PVC)
12. Which type of plastics can be recycled?
    Answer: Thermoplastics
13. What is the name given to plastics that cannot be recycled?
    Answer: Thermosetting plastics
14. Which element is a liquid at ordinary temperature?
    Answer: Mercury
15. What is the chemical name of baking soda?
    Answer: Sodium bicarbonate
16. What is sodium carbonate commonly known as?
    Answer: Washing soda or soda ash
17. What is sodium chloride?
    Answer: Table salt
18. What is table sugar?
    Answer: Cane sugar (Sucrose)
19. What is the element present on lead pencils?
    Answer: Carbon (graphite)
20. Which gas is found in soda water?
    Answer: Carbon dioxide
21. What is the solvent of gold?
    Answer: Aquaregia
22. Name the chemical assed to make tooth pastes white?
    Answer: Titanium dioxide
23. Which chemical is the reason behind the brown colour of human faces?
    Answer: Bilirubin
24. Which chemical is called Chile Salt peter?
    Answer: Sodium Nitrate
25. What is marsh gas?
    Answer: Methane

26. What is the chemical name of Chinese salt?
    Answer: Monosodium glutamate
27. Which is the purest form of carbon?
    Answer: Graphite
28. Which is called white poison?
    Answer: Sugar
29. Which is the simplest sugar?
    Answer: Glucose
30. What is the biological polymer in paper?
    Answer: Cellulose
31. What is the chemical name of Vitamin E?
    Answer: Alpha tocopherol
32. Which Vitamin is also known as retinol?
    Answer: Vitamin A
33. What is Calciferol?
    Answer: Vitamin D
34. What is the other name of thiamine?
    Answer: Vitamin B1
35. Which Vitamin is called niacin?
    Answer: Vitamin B3
36. Which element is excreted through human sweat?
    Answer: Sulphur
37. Pollution of water bodies by which element causes Minamata disease?
    Answer: Mercury
38. Accumulation of which chemical leads to kidney stones?
    Answer: Calcium Oxalate
39. Which heavy metal causes **Itai Itai** disease?
    Answer: Cadmium
40. Which chemical was used earlier for giving anesthesia to patients?
    Answer: Diethyl ether

Diseases spread and agents

S. No.	Diseases	Spread through	Agents
1.	Cold, influenza virus	Air	Virus
2.	Cholera, gastroenteritis and typhoid	Food and water	Bacteria
3.	Ringworm	Direct contact	Fungus
4.	STDs: Syphilis, gonorrhea	Sexual contact	Bacteria
5.	AIDS		Virus(retroviruses)
6.	Malaria	Mosquitoes	Protozoan
7.	Measles, rabies, dengue, hepatitis, chickenpox and cancer	Mosquitoes	Virus
	a. Hepatitis	Water	Virus
	b. Dengue	Mosquitoes	Virus
	c. Chickenpox	air	Virus
8.	Dysentery	Water	Protozoan
9.	Kala-azar	-	Protozoan
10.	COVID-19	Air, Water drop	Novel CORONA virus

(STDs: Sexually Transmitted Diseases,
AIDS: Acquired Immune Deficiency Syndrome)

UNIT - I
OBJECTIVE TYPE AGRICULTURE

1. **The Himalayan rivers are-**
   (a) monsoon fed (b) snow fed (c)ephimeral (d) seasonal
2. **From which of the following states does the river Cauvery originate?**
   (a) Andhra Pradesh (b) Tamil Nadu
   (c) Karnataka (d) Madhya Pradesh
3. **The river basin which is called "Ruhr of India" is-**
   (a) Damodar (b) Hooghly
   (c) Godavari (d) Swarnarekha
4. **The Siachin glacier is in-**
   (a) Jammu (b) Aksaichin (c) Ladakh (d) Garhwal
5. **In which state Maikal range is situated ?**
   (a) Jammu & Kashmir (b) Rajasthan
   (c) Gujarat (d) Madhya Pradesh
6. **Maharashtra plataeu is made up of-**
   (a) alluvial soil (b) coral reef (c) sand stone (d) lava
7. **The widest gap across the western Ghats is-**
   (a) Bhor Ghat (b) Thal Ghat
   (c) Khandwa Gap (d) Palghat
8. **On which of the following states retreating monsoon has more effect?**
   (a) Orissa (b) West Bengal
   (c) Punjab (d) Tamil Nadu
9. **The state in India having dry season for only three or four months every year is-**
   (a) West Bengal (b) Mizoram
   (c) Himachal Pradesh (d) Kerala
10. **North-East monsoon brings maximum rainfall to-**
    (a) Mumbai (b) Delhi (c) Kolkata (d) Chennai

11. The driest place in India is-
(a) Leh (b) Banner (c) Jaisalmer (d) Bikaner

12. Which of the following states touches the boundary four nations ?
(a) Assam (b) Uttar Pradesh
(c) Nagaland (d) Jammu and Kashmir

13. Which of the following lattitudes passes through India?
(a) Equator (b) Tropic of cancer
(c) Tropic of Capricorn (d) Arctic circle

14. The India-Pakistan boundary was settled by-
(a) McMohan line (b) Radcliff line
(c) Durand line (d) Maginot line

15. Which of the following Indian states is land locked ?
(a) Andhra Pradesh (b) Bihar
(c) West Bengal (d) Kerala

16. The southernmost tip of India is-
(a) Lakshadweep (b) Andaman islands
(c) Kanyakumari (d) Rameshwaram

17. Which is the important red soil area in India ?
(a) Indo-Gangetic plain (b) Southern Peninsula
(c) Gujarat and Maharashtra (d) Jammu & Kashmir

18. Which of the following soils is very hard to cultivate ?
(a) Alluvial (b) Red (c) Black (d) Sandy

19. Which area of Madhya Pradesh is known for white tigers?
(a) Rohilkhand (b) Baghel khand
(c) Malwa (d) None of these

20. Telgu Ganga Project is planned to supply drinking water to-
(a) Hyderabad (b) Coimbatore
(c) Chennai (d) Bangalore

21. Thein Dam is being constructed on the river-
(a) Satluj (b) Ravi
(c) Beas (d) None of these

22. Dry farming in India is extensively practised in-
(a) Kanara plains (b) Deccan region
(c) Coromandal plains (d) Punjab plains

23. In which state of India wet agriculture is practised?
(a) Tamil Nadu (b) Karnataka
(c) Andhra Pradesh (d) Kerala

24. Jhum cultivation is prevalent in-
(a) Punjab (b) Haryana
(c) Madhya Pradesh (d) Manipur

25. The highest milk producing breed of goat in India is-
(a) Barbari (b) Beetal
(c) Jamnapari (d) Black Bengal

26. After u. p. which state leads in production of sugarcane ?
(a) Tamil Nadu (b) Maharashtra
(c) Bihar (d) Andhra Pradesh

27. In which of the following states does the maximum density of canals lie?
(a) Tamil Nadu (b) Andhra Pradesh
(b) Bihar (d) U. P.

28. Mathura refinery receives crude oil from-
(a) Cauvery basin (b) Vadodara
(c) Kandla (d) None of these

29. In which of the following states Koyali having a petroleum refinery is located ?
(a) Bihar (b) Assam (c) Gujarat (d) U. P.

30. Lignite is found in India in-
(a) Rourkela (b) Hazaribagh
(c) Raniganj (d) Neyveli

31. The state in south India with a thermal station at Tuticoran is-
(a) Karnataka (b) Kerala
(c) Tamil Nadu (d) Andhra Pradesh

32. At present, India's largest mineral resource is-
(a) copper (b) coal
(c) iron-ore (d) none of these

33. The only place where tin is found in India is-
(a) Hazaribagh (b) Rewa (c) Salem (d) Quilon

34. India has the maximum amount of foreign trade with-
(a) U. S. A. (b) China
(c) Japan (d) Germany

35. Nhava Sheva, one of India's major ports is in-
(a) Gujarat (b) West Bengal
(c) Maharashtra (d) Andhra Pradesh

36. Which of the following ports does not have a natural harbour?
(a) Cochin (b) Mumbai
(c) Chennai (d) Vishakhapatnam

37. The first Electric train in India (Deccan Queen) was run between-
(a) Mumbai and Surat
(b) Howrah and Delhi
(c) Kalyan and Pune
(d) New Delhi and Chennai

38. Gitanjali Express runs between-
(a) Howrah-Puri (b) Howrah-Jammu
(c) Howrah-Mumbai (d) Howrah-Jaipur

39. The first shore-based steel plant in India is in-
(a) Haldia (b) Kandla
(c) Paradeep (d) Vishakhapatnam

40. The place 'Avadi' which is known for the manufacture of Vijay anta tanks for the Indian Army is located in-
(a) Orissa (b) Karnataka
(c) Andhra Pradesh (d) Tamil Nadu

41. In which industry is sal wood used mostly?
(a) Railway sleepers (b) Matches
(c) Paper (d) None of these

42. The main centre of diamond cutting industry in India is-
(a) Panna (b) Kolar (c) Surat (d) Agra

43. Where is the Hindustan Shipyard located?
(a) Cochin (b) Vishakhapatnam
(c) Mumbai (d) Kolkata

44. The satellite freight city is being developed near which of the following cities?
(a) Kolkata (b) Bangalore (c) Kandla (d) Gurgaon

45. India's oldest Planetorium is located at-
(a) Kolkata (b) Mumbai
(c) Patna (d) Ahmadabad

46. The first submarine built by India is-
(a) INS Shalki (b) INS Ganga
(c) INS Godavari (d) INS Himgiri

47. Vijayawada is situated on the bank of the river--
(a) Tungabhadra (b) Godavari
(c) Krishna (d) Cauvery

48. NEFA is the old name of-
(a) Karnataka (b) Tamil Nadu
(c) Nagaland (d) Arunachal Pradesh

49. Kaiga in Karnataka is famous for-
(a) nuclear plant (b) ancient fort
(c) fertiliser plant (d) cement plant

50. Palk straits separates India from-
(a) Bangladesh (b) Pakistan
(c) Sri Lanka (d) Maldives

Answer Keys (Objective Type Agriculture - Unit I)

1. (b)	2. (c)	3. (a)	4. (c)	5. (d)	6. (d)	7. (d)	8. (d)	9. (d)	10. (d)
11. (c)	12. (d)	13. (b)	14. (b)	15. (b)	16. (c)	17. (d).	18. (b)	19. (b)	20. (c)
21. (b)	22. (b)	23. (d)	24. (d)	25. (b)	26. (b)	27. (d)	28. (c)	29. (c)	30. (d)
31. (c)	32. (b)	33. (b)	34. (a)	35. (c)	36. (c)	37. (c)	38. (c)	39. (d)	40. (d)
41. (a)	42. (c)	43. (b)	44. (d)	45. (a)	46. (a)	47. (c)	48. (d)	49. (a)	50. (c)

GENERAL SCIENCE
OBJECTIVE TYPE – I

1. Brass gets discoloured in air because of the presence of which of the following gases in air?
   A. Oxygen B. Hydrogen sulphide
   C. Carbon dioxide D. Nitrogen
   Answer: Hydrogen sulphide
2. Which of the following is a non metal that remains liquid at room temperature?
   A. Phosphorous B. Bromine
   C. Chlorine D. Helium
   Answer: Bromine
3. Chlorophyll is a naturally occurring chelate compound in which central metal is
   A. copper B. magnesium
   C. iron D. calcium
   Answer: Magnesium
4. Which of the following is used in pencils?
   A. Graphite B. Silicon
   C. Charcoal D. Phosphorous
   Answer: Graphite
5. Which of the following metals forms an amalgam with other metals?
   A. Tin B. Mercury C. Lead D. Zinc
   Answer: Mercury
6. Chemical formula for water is
   A. $NaAlO_2$ B. H_2O C. Al_2O_3 D. $CaSiO_3$
   Answer: H_2O
7. The gas usually filled in the electric bulb is
   A. Nitrogen B. Hydrogen
   C. Carbon dioxide D. Oxygen
   Answer: Nitrogen

8. Washing soda is the common name for
   A. Sodium carbonate B. Calcium bicarbonate
   C. Sodium bicarbonate D. Calcium carbonate
   Answer: Sodium carbonate
9. Quartz crystals normally used in quartz clocks etc. is chemically
   A. Silicon dioxide
   B. Germanium oxide
   C. A mixture of germanium oxide and silicon dioxide
   D. Sodium silicate
   Answer: Silicon dioxide
10. Which of the gas is not known as green house gas?
    A. Methane B. Nitrous oxide
    C. Carbon dioxide D. Hydrogen
    Answer: Hydrogen
11. The hardest substance available on earth is
    A. Gold B. Iron C. Diamond D. Platinum
    Answer: Diamond
12. The variety of coal in which the deposit contains recognizable traces of the original plant material is
    A. bitumen B. anthracite C. lignite D. peat
    Answer: Peat
13. Tetraethyl lead is used as
    A. Pain killer B. Fire extinguisher
    C. Mosquito repellenet D. Petrol additive
    Answer: Petrol aditive
14. The inert gas which is substituted for nitrogen in the air used by deep sea divers for breathing, is
    A. Argon B. Xenon C. Helium D. Krypton
    Answer: Helium
15. The gases used in different types of welding would include
    A. Oxygen and hydrogen

B. Oxygen, hydrogen, acetylene and nitrogen
C. Oxygen, acetylene and argon
D. Oxygen and acetylene
Answer: Oxygen and acetylene

16. In which of the following activities silicon carbide is used?
A. Making cement and glass
B. Disinfecting water of ponds
C. cutting very hard substances
D. Making casts for statues
Answer: cutting very hard substances

17. Which of the following has an organic origin?
A. Nylon B. Sand C. Coal D. Bakelite
Answer: Coal

18. What is the name given to molten rocks erupted by a volcano?
A. Firestone B. Crater C. Lava D. Geyser
Answer: Lava

19. Which planet is known as the Red Planet?
A. Earth B. Mars C. Mercury D. Jupiter
Answer: Mars

20. Where is the Sea of Tranquility?
A. Near the Bermuda Triangle
B. On the Moon
C. In the Pacific Ocean
D. In the Atlantic Ocean
Answer: On the Moon

21. Which of the following is necessary for burning (combustion)?
A. Nitrogen B. Oxygen C. Carbon D. Petrol
Answer: Oxygen

22. Which gas in the atmosphere protects us from harmful ultraviolet radiations?
A. Oxygen B. Nitrogen
C. Ozone D. Carbon Dioxide
Answer: Ozone

23. The animal kingdom is formally classified into major groups known as
    A. Sub Phyla B. Orders C. Phyla D. Families
    Answer: Phyla
24. Enzymes and anti-bodies are mainly made of
    A. Proteins B. Vitamins
    C. Carbohydrates D. Fats
    Answer: Proteins
25. The atom is made of
    A. Positrons, Neutrons and Electrons
    B. Protons and Quarks
    C. Neutrinos, Gamma Rays and Positrons
    D. Protons, Neutrons and Electrons
    Answer: Protons, Neutrons and Electrons
26. Which rare element would you associate with Marie and Pierre Curie?
    A. Platinum B. Radium C. Gold D. Uranium
    Answer: Radium
27. Who discovered Penicillin?
    A. Louis Pasteur B. Sir Alexander Fleming
    C. Watson and Crick D. Robert Koch
    Answer: Sir Alexander Fleming
28. Who invented the miner's safety-lamp?
    A. Thomas Alva Edison B. Alexander Graham Bell
    C. Sir Humphrey Davy D. Sir Frank Whittle
    Answer: Sir Humphrey Davy
29. The Sun is a
    A. Star B. Huge Planet
    C. Satellite D. Comet
    Answer: Star
30. What is known as the universal solvent?
    A. Sulphuric Acid B. Water
    C. Hydrochloric Acid D. Aqua Regia
    Answer: Water

31. Who discovered the circulation of blood?
    A. Hargobind Khorana   B. William Harvey
    C. Louis Pasteur   D. Edward Jenner
    Answer: William Harvey
32. A communicable disease can be passed on from one person to another.
    A. True   B. False
    Answer: True
33. The three states of matter are solid, liquid and gas.
    A. True   B. False
    Answer: True
34. The science that deals with the study of sound is known as acoustics.
    A. True   B. False
    Answer: True
35. The tail of a comet always points towards the Sun.
    A. True   B. False
    Answer: False
36. Who wrote the book "The Origin of Species"?
    A. Charles Darwin   B. Stephen Hawking
    C. Sir Alexander Fleming   D. Louis Pasteur
    Answer: Charles Darwin
37. Which gas evolves when charcoal is burnt?
    A. Nitrogen   B. Carbon Dioxide
    C. Ozone   D. Oxygen
    Answer: Carbon Dioxide
38. The Earth is surrounded by an insulating blanket of gases which protects it from the light and heat of the Sun. This insulating layer is called the
    A. Lithosphere   B. Hydrosphere
    C. Atmosphere   D. Biosphere
    Answer: Atmosphere
39. What is the total number of bones in the human body?
    A. 32   B. 196   C. 206   D. 512
    Answer: 206

40. The human skeleton consists of 206 bones (babies have over 300). Bones manufacture red blood cells and store calcium. How many planets are there in our solar system?

    A. 8　　B. 9　　C. 10　　D. 12

    Answer: 9

GENERAL SCIENCE OBJECTIVE TYPE - II

1. Which of the following is the lightest metal ?
   (A) Mercury (B) Silver (C) Lithium (D) Lead
2. The most important ore of Aluminium is –
   (A) Bauxite (B) Calamine
   (C) Calcite (D) Galena
3. The element present in the largest amount in rocks and minerals is –
   (A) Carbon (B) Silicon
   (C) Hydrogen (D) Aluminium
4. An alloy used in making heating elements for electric heating devices is –
   (A) Solder (B) Alloysteel
   (C) Nichrome (D) German Silver
5. German Silver is an alloy of –
   (A) Copper, Silver & Nickel
   (B) Silver, Copper & Aluminium
   (C) Zinc, Copper & Nickel
   (D) Silver, Zinc & Nickel
6. Air is a/an –
   (A) Compound (B) Element
   (C) Mixture (D) Electrolyte
7. Which of the following is the best conductor of Electricity?
   (A) Ordinary water (B) Sea water
   (C) Boiled water (D) Distilled water
8. Balloons are filled with –
   (A) Helium (B) Oxygen (C) Nitrogen (D) Argon
9. The charcoal used to decolourise raw sugar is –
   (A) Animal charcoal (B) Sugar charcoal
   (C) Cocoanut charcoal (D) Wood charcoal

10. The most abundant metal in the earth's crust is –
    (A) Zinc (B) Copper
    (C) Aluminium (D) Iron
11. The gas used to extinguish fire is –
    (A) Neon (B) Nitrogen
    (C) Carbon dioxide (D) Carbon Monoxide
12. In which of the following activities Silicon Carbide is used?
    (A) Making cement and glass
    (B) Disinfecting water and ponds
    (C) Making castes for statues
    (D) Cutting very hard substances
13. The element common to all acids is –
    (A) Oxygen (B) Hydrogen (C) Nitrogen (D) Sulphur
14. Gobar gas contains mainly –
    (A) Methane (B) Carbon dioxide
    (C) Butane (D) Carbon Monoxide
15. The most malleable metal is –
    (A) Silver (B) Gold
    (C) Aluminium (D) Sodium
16. Tetraethyl lead is used as –
    (A) Mosquito repellent (B) Pain Killer
    (C) Fire extinguisher (D) Petrol additive
17. Which of the following is used in beauty parlours for hair setting ?
    (A) Phosphorus (B) Sulphur (C) Chlorine (D) Silicon
18. Which of the following is a non-metal that remains liquid in room temperature ?
    (A) Bromine (B) Chlorine
    (C) Helium (D) Phosphorus
19. From which mineral is radium obtained –
    (A) Limestone (B) Haematitie
    (C) Pitchblende (D) Rutile

20. What is laughing gas ?
    (A) Carbon dioxide (B) Sulphur dioxide
    (C) Nitrogen Dioxide (D) Nitrous oxide
21. Which of the following is used as a material for making protective windows in space probes ?
    (A) Steel (B) Graphite
    (C) Tungsten (D) Diamond
22. Which form of phosphorus is used in safety metals?
    (A) Red Phosphorus (B) White Phosphorus
    (C) Yellow Phosphorus (D) Black Phosphorous
23. Stainless steel is an alloy of –
    (A) Iron, Carbon and Nickel
    (B) Iron and Manganese
    (C) Iron, Chromium and Zinc
    (D) Iron, Chromium and Nickel
24. Which of the following is not correctly matched ?
    (A) Galena : Lead Sulphide
    (B) Green vitriol : Copper sulphate
    (C) Plaster of Paris : Calcium sulphate
    (D) Calomel : Mercurous Chloride
25. Which of the following metals forms an amalgum with other metals ?
    (A) Lead (B) Tin (C) Zinc (D) Mercury
26. Which of the following is not an ore of Aluminium?
    (A) Felsper (B) Bauxite (C) Cryolite (D) Azurite
27. Most soluble in water is –
    (A) Camphor (B) Sugar
    (C) Sulphur (D) Common Salt
28. Bromine is –
    (A) Colourless gas insoluble in water
    (B) A highly inflammable gas
    (C) A Black solid.
    (D) A red liquid.

29. The metal used in storage batteries –
(A) Iron (B) Copper (C) Lead (D) Zinc

30. Water has maximum density at –
(A) –4°C (B) 0°C (C) 4°C (D) 100°C

31. The chemial name of Uria is –
(A) Aneurin (B) Chloroetane
(C) Carbamide (D) None of these

32. Permanent hardness of water can be removed by adding –
(A) Potassium Permanganate (B) Chlorine
(C) Bleaching Powder (D) Washing Soda

33. The formula of Plaster of Paris is –
(A) $CaSO_4$ (B) $CaSO_4, 2H_2O$
(C) $2CaSO_4, 4H_2O$ (D) $2CaSO_4, H_2O$

34. Which of the following substance is a bad conductor of electricity but a good conductor of heat?
(A) Asbestos (B) Celluloid (C) Purspecks (D) Mica

35. Carborandum is –
(A) Silicon Dioxide (B) Silicon Carbide
(C) Silicon Nitride (D) Silicon Phosphide

36. Which of the following is the best conductor of electricity ?
(A) Copper (B) Mica (C) Zinc (D) Silver

37. Washing soda is the common name of –
(A) Calcium Carbonate
(B) Calcium Bi-Carbonate
(C) Sodium Carbonate
(D) Sodium Bi-Carbonate

38. The filament of electric bulb is made of –
(A) Iron (B) Nichrome
(C) Tungsten (D) Graphite

39. Bleaching powder is made from –
(A) Sulphur dioxide and gypsum
(B) Chlorine and Charcoal

(C) Soda ash and lime
(D) Lime and Chlorine

40. The two elements that are frequently used for making transistors are –
(A) Boron and Aluminium
(B) Silicon and Germenium
(C) Iridium and Tungsten
(D) Niobium and Columbium

41. Which of the following gas is not known as green house gas ?
(A) Methane (B) Carbon dioxide
(C) Nitrous oxide (D) CFC

42. Which of the following is not a neutral oxide ?
(A) Carbon Monoxide (B) Sulphur Dioxide
(C) Water (D) Nitric Oxide

43. Which of the following is used in making smoke bombs ?
(A) Sulphur (B) Phosphorus
(C) Hydrogen (D) Carbon

44. Caustic Soda is –
(A) Nacl (B) Na_2CO_3
(C) NaOH (D) $NaHCO_3$

45. Chemical formula of Water glass is –
(A) Na_2SiO_3 (B) Al_2O_3 (C) $NaAlO_2$ (D) $CaSiO_3$

46. Which of the following is not a bleaching agent?
(A) Sulpher di-oxide (B) Carbon di-oxide
(C) Sodium hypochlorite (D) Chlorine

47. The gas usually filled in electric bulb is –
(A) Hydrogen (B) Oxygen
(C) Nitrogen (D) Carbon di-oxide

48. Heavy water is –
(A) Tritium oxide (B) Deuterium oxide
(C) Rain water (D) Water at 4°C

49. The gases used in different types of welding would include
    (A) Oxygen & Hydrogen
    (B) Oxygen, acetylene & argon
    (C) Oxygen & acetylene
    (D) Oxygen, hydrogen & nitrogen
50. Which of the following alloys is used for making magnets?
    (A) Duralumin (B) Stainless Steel
    (C) Alnico (D) Magnelium

Answer Keys: (General Science (Objective Type) - II

1. (c)	2. (a)	3. (b)	4. (c)	5. (c)	6. (c)	7. (b)	8. (a)	9. (d)	10. (c)
11. (c)	12. (d)	13. (b)	14. (a)	15. (b)	16. (d)	17. (b).	18. (a)	19. (c)	20. (d)
21. (d)	22. (a)	23. (d)	24. (b)	25. (d)	26. (d)	27. (b)	28. (d)	29. (c)	30. (a)
31. (c)	32. (d)	33. (d)	34. (d)	35. (b)	36. (d)	37. (c)	38. (c)	39. (d)	40. (b)
41. (c)	42. (b)	43. (b)	44. (c)	45. (a)	46. (b)	47. (c)	48. (b)	49. (c)	50. (c)

GENERAL SCIENCE
OBJECTIVE TYPE - III

1. Permanent hardness of water may be removed by the addition of –
   (A) Alum
   (B) Sodium Carbonate
   (C) Lime
   (D) Potassium Permanganate
2. Potassium nitrate is used in –
   (A) Salt (B) Glass
   (C) Fertilizer (D) Medicine
3. In which type of rocks are metals like Gold and Copper mostly found ?
   (A) Old igneous (B) Old sedimentary
   (C) Old metamorphic (D) New metamorphic
4. Which of the following gases is lighter than air?
   (A) Carbon dioxide (B) Oxygen
   (C) Ammonia (D) Chlorine
5. Which of the following is used as lubricant ?
   (A) Graphite (B) Silica
   (C) Diamond (D) Iron Oxide
6. The purest form of iron is –
   (A) Steel (B) Pig iron
   (C) Cast iron (D) Wrought iron
7. The percentage of carbon is the least in –
   (A) Grey cast iron (B) Wrought iron
   (C) White cast iron (D) Steel
8. The colour of Emerald is –
   (A) Violet (B) Yellow
   (C) Deep Green (D) Prussian Blue
9. Permanent hardness of water, due to sulphates of the metal, can be destroyed by the use of –

(A) Nitrates (B) Zeolites
(C) Sulphonates (D) None of these

10. Rust is –
   (A) A mixture of Fe_2O_3 and $Fe(OH)_2$
   (B) A mixture of FeO and $Fe(OH)_2$
   (C) FeO only
   (D) A mixture of Fe_2O_3, $3H_2O$ and FeO
11. Liquefied Petroleum gas (LPG) consists of mainly –
   (A) Methane, Ethane and Hexane
   (B) Ethane, Hexane and Nonane
   (C) Methane, Butane and Propane
   (D) Methane, Hexane and Nonane
12. The metal that is present in Photo Films is –
   (A) Mercury (B) Platinium
   (C) Magnesium (D) Silver
13. Which of the following is in liquid form at room temperature?
   (A) Cerium (B) Sodium
   (C) Francium (D) Lithium
14. Soda water contains –
   (A) Nitrous acid (B) Carbonic acid
   (C) Carbon dioxide (D) Sulphuric acid
15. Which of the following is basis of the modern periodic table?
   (A) Atomic mass (B) Atomic number
   (C) Atomic size (D) Atomic volume
16. Of the following metals, which one pollutes the air of a big city ?
   (A) Copper (B) Cadmium
   (C) Lead (D) Chromium
17. Brass is an alloy of –
   (A) Nickel and Copper (B) Copper and Silver
   (C) Nickel and Zinc (D) Zinc and Copper

18. The element required for Solar energy conversion –
    (A) Beryllium (B) Silicon
    (C) Tantalum (D) Ultra pure carbon
19. Monazite is an ore of –
    (A) Sodium (B) Titanium
    (C) Thorium (D) Zirconium
20. Potassium Permanganate is used for purifying drinking water, because –
    (A) It dissolves the impurities of water
    (B) It is a sterilizing agent
    (C) It is an oxidising agent
    (D) It is a reducing agent.
21. The presence of which of the following salts in water causes corrosion in steam boilers ?
    (A) Sodium Chloride
    (B) Magnesium Chloride
    (C) Calcium bicarbonate
    (D) Potassium bicarbonate
22. Water is a good solvent of ionic salts because –
    (A) It has no colour
    (B) It has a boiling point
    (C) It has a high dipole moment
    (D) It has a high specific heat
23. The hardest substance available in earth is –
    (A) Platinum (B) Silicon (C) Diamond (D) Gold
24. Which of the following is the best conductor of electricity?
    (A) Zinc (B) Copper (C) Gold (D) Silver
25. The ratio of pure gold in 18 carat gold is –
    (A) 60% (B) 75% (C) 80% (D) 100%
26. Cow milk is a rich source of –
    (A) Vitamin A (B) Vitamin B_1
    (C) Vitamin C (D) Vitamin D

27. The element found in the surface of the Moon is –
    (A) Tin (B) Tungsten
    (C) Tantalum (D) Titanium
28. The average salinity of sea water is –
    (A) 2% (B) 3% (C) 2.5% (D) 3.5%
29. Which of the following is the most stable element?
    (A) Uranium (B) Hydrogen (C) Oxygen (D) Lead
30. Find the odd one ?
    (A) Marble (B) Chalk
    (C) Lime Stone (D) Slaked Lime
31. Diamond is harder than graphite because of –
    (A) Difference of layers of atoms
    (B) Tetrahedral structure of diamonds
    (C) Difference of crystaline structures
    (D) None of these
32. A super conductor is characterised by –
    (A) Zero permeability (B) Low permeability
    (C) High permeability (D) Infinite permeability
33. Of the various forms of carbon, which one is used for absorption purpose ?
    (A) Diamond (B) Graphite
    (C) Coke (D) Wood Charcoal
34. The main constituents of Pearl are –
    (A) Calcium Carbonate, Magnesium Carbonate
    (B) Calcium oxide, Ammonium Chloride
    (C) Argonite, Chonchiolin
    (D) Ammonium Sulphate, Sodium Carbonate
35. An element that does not occur in nature but can be produced artificially is –
    (A) Thorium (B) Radium
    (C) Plutonium (D) Uranium
36. The inert gas which is substituted for nitrogen in the air, used by deep sea drivers for breathing is –
    (A) Helium (B) Argon (C) Krypton (D) Xenon

37. The group of metals Fe, Co, Ni may be best called as –
(A) Alkali metals (B) Rare metals
(C) Main group metals (D) Transition metals

38. Which of the following characters is not shown by hydrogen –
(A) It burns in air to form water
(B) It supports combustion
(C) It combines with almost all metals forming hydrides
(D) It readily combines with fluorine and chlorine

39. Which of the following elements is obtained from sea weeds ?
(A) Argon (B) Sulphur (C) Vanadium (D) Iodine

40. The metallic constituents of hard water are –
(A) Magnesium, Calcium and tin
(B) Iron, tin and calcium
(C) Calcium, magnesium and iron
(D) Magnesium, tin and iron

41. The property of a substance to absorb moisture from the air on exposure is called –
(A) Osmosis (B) Desiccation
(C) Efflorescence (D) Deliquescene

42. Brass gets discoloured in air due to the presence of which gas in air –
(A) Carbon dioxide (B) Oxygen
(C) Hydrogen Sulphide (D) Nitrogen

43. Quartz crystals normally used in quartz clocks etc. is chemically –
(A) Sodium Silicate (B) Silicon dioxide
(C) Germenium dioxide (D) A mixture of (B) & (C)

44. Bell Metal is an alloy of –
(A) Brass and Nickel (B) Zinc and Copper
(C) Tin and Copper (D) Nickel and Copper

45. The high temperature superconductors are –
    (A) Metal alloys (B) Ceramic oxides
    (C) Inorganic polymers (D) Pure rare earth metals
46. The ingredients of Gun metal are –
    (A) Iron, tin (B) Copper, tin
    (C) Iron, brass, tin (D) Iron, zinc, titanium
47. The gas usually causing explosions in coal mines is –
    (A) Hydrogen (B) Carbon monoxide
    (C) Air (D) Methane
48. Cotton fibers are made of –
    (A) cellulose (B) starch (C) proteins (D) fats
49. Which of the following ores does not contain iron?
    (A) Haematite (B) Magnetite
    (C) Limonite (D) Cassiterite
50. Nuclear fission is caused by the impact of –
    (A) Proton (B) Electron
    (C) Neutron (D) (A) & (b)

Answer Keys (General Science (Objective Type) - III

1. (b)	2. (b)	3. (a)	4. (c)	5. (a)	6. (d)	7. (d)	8. (c)	9. (b)	10. (d)
11. (c)	12. (d)	13. (c)	14. (c)	15. (b)	16. (c)	17. (d).	18. (b)	19. (c)	20. (c)
21. (b)	22. (c)	23. (c)	24. (d)	25. (b)	26. (b)	27. (d)	28. (d)	29. (d)	30. (d)
31. (c)	32. (a)	33. (d)	34. (a)	35. (c)	36. (a)	37. (d)	38. (b)	39. (d)	40. (c)
41. (d)	42. (c)	43. (b)	44. (c)	45. (b)	46. (b)	47. (d)	48. (a)	49. (d)	50. (c)

GENERAL SCIENCE
OBJECTIVE TYPE – IV

1. Which is a good nuclear fuel ?
   (A) Uranium – 238 (B) Neptunium – 239
   (C) Thorium – 236 (D) Plutonium - 236
2. Long distance photography is facilitated by –
   (A) Visible light (B) X-rays
   (C) Infra Red rays (D) Ultra violet rays
3. Which radioactive pollutant has recently drawn the attention of the public due to its occurrence in the building materials ?
   (A) Radium (B) Radon
   (C) Thorium (D) Plutonium
4. In an atomic explosion enormous energy is released which is due to the –
   (A) Conversions of neutrons into protons
   (B) Conversion of chemical energy into nuclear energy
   (C) Conversion of mechanical energy into nuclear energy
   (D) Conversion of mass into energy
5. Which of the following is used as a coolant in nuclear reactors?
   (A) Heavy water (B) Cadmium
   (C) Liquid sodium (D) Graphite
6. The isotope of Uranium capable of sustaining chain reaction is –
   (A) U 235 (B) U 238
   (C) U 239 (D) None of these
7. Radioactive disintegration of Uranium ultimately results in formation of –
   (A) Radium (B) Thorium (C) Polonium (D) Lead

8. Atom bomb is made on the basis of –
   (A) Nuclear fusion (B) Nuclear fission
   (C) Both (A) & (B) (D) None of these
9. Which of the following has least penetrating power?
   (A) Alpha particles (B) Beta particles
   (C) Gamma particles (D) All have same power
10. What is increasing order of the wave lengths of the following colours –
   1. Orange 2. Indigo 3. Yellow 4. Violet
   (A) 1, 2, 3, 4 (B) 3, 4, 1, 2
   (C) 4, 3, 2, 1 (D) 4, 2, 3, 1
11. Atoms are composed of –
   (A) Electrons only (B) Protons only
   (C) Electrons & Protons (D) Electrons & Nuclei
12. The age of most ancient geological formation is estimated by –
   (A) Potassium-Argon method (B) Ra-Si method
   (C) C14 method (D) Uranium-lead method
13. Which of the following is used as a moderator in nuclear reactor ?
   (A) Ordinary water (B) Radium
   (C) Thorium (D) Graphite
14. Isotopes are separated by –
   (A) Filtration (B) Sublimation
   (C) Distillation (D) Crystalization
15. Who suggested that the most of the mass of atom is located in nucleus ?
   (A) Bohr (B) Thomson
   (C) Rutherford (D) Avogadro
16. Meson particles are found in –
   (A) R-rays (B) X-rays
   (C) Laser beam (D) Cosmic rays

17. An atom of an element with mass number 23 and atomic number 11 will have –
    (A) 11 neutrons, 12 protons and 11 electrons
    (B) 11 protons, 12 neutrons and 11 electrons
    (C) 11 protons, 12 electrons and 11 neutrons
    (D) 23 protons and 11 electrons
18. The location and energy of an electron in an atom can be specified by –
    (A) Atomic mass (B) Atomic number
    (C) Quantum numbers (D) None of these
19. Old-written material, which cannot be read easily can be read by –
    (A) Cosmic Rays (B) Ultraviolet rays
    (C) Infra Red rays (D) None of these
20. In an atomic nucleus, neutrons and protons are held together by –
    (A) Gravitational forces (B) Magnetic forces
    (C) Exchange forces (D) Coulombic forces
21. Which of the following elements is non-radioactive?
    (A) Uranium (B) Thorium
    (C) Plutonium (D) Zirconium
22. According to Dalton's atomic theory the smallest particle which can exist independently is –
    (A) An atom (B) A molecule
    (C) A cation (D) An anion
23. The recent atomic weight scale is based on –
    (A) 1H1 (B) 1H2 (C) 6C12 (D) 8O16
24. The neutral atom's two isotopes differ in the number of –
    (A) Electron shells (B) Protons
    (C) Valence electrons (D) Neutrons
25. Optical fibres are mainly used in –
    (A) Communication (B) Weaving
    (C) Musical Instruments (D) Food Industry

26. The first synthetic fibre made by man was –
    (A) Rayon (B) Nylon
    (C) Polyester (D) Terycott
27. Which variety of glass is heat resistant ?
    (A) Hard glass (B) Flint glass
    (C) Pyrex glass (D) Bottle glass
28. Which of the following is used for removing air bubbles from glass during its manufacture ?
    (A) Fledspar (B) Arsenic oxide
    (C) Potassium Carbonate (D) Soda Ash
29. What are soaps ?
    (A) Salts of silicates
    (B) Ester of heavy fatty acids
    (C) Sodium or potassium salts of heavier fatty acids
    (D) Mixture of glycerol and alcohol
30. Which of the following is used in making artificial sweetener saccharine ?
    (A) Benzene (B) Toluene
    (C) Phenol (D) Aniline
31. Rayon is chemically –
    (A) Glucose (B) Amylase (C) Cellulose (D) Peptin
32. Which of following is used as a filter in rubber tyres?
    (A) Graphite (B) Coal
    (C) Coke (D) Carbon Black
33. In vulcanization, natural rubber is heated with –
    (A) Carbon (B) Sulphur
    (C) Silicon (D) Phosphorus
34. Which type of fire extinguisher is used for petroleum fire ?
    (A) Foam type (B) Soda acid type
    (C) Powder type (D) None of these
35. Which is/ are the important raw materials in cement industry ?
    (A) Limestone (B) Gypsum & Clay
    (C) Clay (D) Limestone & Clay

36. Which of the following is the petroleum wax?
    (A) Bees wax (B) Carnauba wax
    (C) Paraffin wax (D) Jojoba wax
37. Which of the following fibres is generally preferred for making gauze and lint ?
    (A) Cotton (B) Rayon
    (C) Nylon (D) Terylene
38. Rubber is coagulated from latex by adding –
    (A) Hydrochloric acid (B) Acetic acid
    (C) Sulphuric acid (D) Carbonic acid
39. Paper is manufactured by –
    (A) Wood and resin
    (B) Wood, Sodium and Bleaching powder
    (C) Wood and bleaching powder
    (D) Wood, Calcium, hydrogen sulphate and resin
40. Which synthetic fibre is known as artificial silk ?
    (A) Cotton (B) Rayon (C) Terylene (D) Nylon
41. Deep blue colour is imparted to glass by the presence of –
    (A) Iron oxide (B) Cupric oxide
    (C) Nickel oxide (D) Cobalt oxide
42. The combustible material at the tip of a safety match stick is –
    (A) Sulphur (B) Manganese dioxide
    (C) Phosphorus (D) Antimony sulphide
43. Which of the following roofs provide better protection against fire ?
    (A) Cement slab (B) Asbestos sheet
    (C) Reinforced concrete (D) None of these
44. The material used for bleaching paper pulp is –
    (A) Chlorine (B) Caustic soda
    (C) Sodium Hypochlorite (D) Lime
45. Which of the following is a protein ?
    (A) Wool (B) Starch
    (C) Natural rubber (D) Cellulose

46. Soap is a mixture of sodium or potassium salts of –
    (A) Monocarboxylic acid (B) Glycerols
    (C) Dicarboxylic acid (D) Tricarboxylic acid
47. Which of the following material is used to prepare blue-black ink ?
    (A) Gallic acid (B) Tannic acid
    (C) Ferrous sulphate (D) All of these
48. The type of glass used in making prisms and lenses is –
    (A) Soft glass (B) Pyrex glass
    (C) Jena glass (D) Flint glass
49. Carbon tetrachloride fire extinguisher should not be used in closed room because it produces poisonous glass called –
    (A) Carbon Monoxide (B) Phosphine
    (C) Phosgene (D) None of these
50. Which of the following is commonly called a polyamide ?
    (A) Rayon (B) Orion (C) Terylene (D) Nylon

Answer Keys (General Science Objective Type) - IV

1. (c)	2. (c)	3. (c)	4. (d)	5. (c)	6. (a)	7. (d)	8. (b)	9. (a)	10. (d)
11. (d)	12. (c)	13. (d)	14. (c)	15. (c)	16. (d)	17. (b)	18. (c)	19. (c)	20. (c)
21.	22.	23.	24.	25.	26.	27.	28.	29.	30.
31. (c)	32. (d)	33. (b)	34. (c)	35. (d)	36. (c)	37. (b)	38. (b)	39. (d)	40. (b)
41. (d)	42. (d)	43. (b)	44. (c)	45. (c)	46. (a)	47. (d)	48. (d)	49. (d)	50. (d)

GENERAL SCIENCE OBJECTIVE TYPE – V

1. The major ingredient of leather is –
   (A) Carbohydrate (B) Collagen
   (C) Polymer (D) Nucleic acid
2. Glass is made from the mixture of –
   (A) Quartz and mica (B) Sand and salt
   (C) Sand and silicates (D) None of these
3. Epoxy resins is used as –
   (A) Moth repellants (B) Insecticides
   (C) Detergents (D) Adhesives
4. Polythene is industrially prepared by the polymerization of –
   (A) Methane (B) Styrene
   (C) Acetylene (D) Ethylene
5. A mixture of water and alcohol can be separated by –
   (A) Filtration (B) Evaporation
   (C) Distillation (D) Decantation
6. A substance which changes readily into vapour without heating is called –
   (A) Efflorescent (B) Synthetic
   (C) Volatile (D) Effervescent
7. In which following processes light energy is converted into chemical energy ?
   (A) Respiration (B) Fermentation
   (C) Photosynthesis (D) Photorespiration
8. Cooking oil can be converted into vegetables ghee by the process of –
   (A) Oxidation (B) Hydrogenation
   (C) Distillation (D) Crystallisation

9. Photosynthesis is –
   (A) An exothermic process
   (B) An endothermic process
   (C) A neutral process
   (D) A thermostatic process
10. Which of the following substances exhibit the property of sublimation ?
    (A) Ice (B) Wax
    (C) Camphor (D) Ethyl Alcohol
11. Combustion is the process in which –
    (A) Heat is produced (B) Light is produced
    (C) Heat and Light is produced (D) None of these
12. The chemical used as a fixer in photography is –
    (A) Borax
    (B) Sodium thiosulphate
    (C) Sodium Sulphate
    (D) Ammonium persulphate
13. A mixture of iron filings and sand can be separated by –
    (A) Heating (B) Sublimation
    (C) Hand picking (D) Magnetic Separation
14. In which of the following process, Vanadium Pentoxide is used as a catalyst ?
    (A) Contact process (B) Haber process
    (C) Solvary process (D) None of these
15. The rate of chemical reaction does not depend on –
    (A) Concentration (B) Catalyst
    (C) Temperature (D) Pressure
16. Which of the following metals can displace hydrogen from dilute acids ?
    (A) Zinc (B) Gold (C) Copper (D) Silver
17. Reaction of alcohol, with carboxylic acid is known as –
    (A) Substitution reaction (B) Addition reaction
    (C) Esterification (D) Hydrogenation

18. Saponification involves the hydrolysis of fats and oils by –
    (A) Water (B) Washing soda
    (C) Stearic acid (D) Caustic acid
19. Which of the following gases is obtained by the reaction of water with Calcium Carbide?
    (A) Acetylene (B) Ethylene
    (C) Methane (D) Water gas
20. The gas liberated during the reaction of copper with dilute nitric acid is –
    (A) NO_2 (B) N_2O_5 (C) O_2 (D) NO
21. Which of following substances undergoes chemical change on heating ?
    (A) Sodium Chloride (B) Silica
    (C) Lead Nitrate (D) Platinum wire
22. Water is neither acidic nor alkaline because –
    (A) it cannot accept or donate protons
    (B) it boils at high temperature
    (C) it can dissociate into equal number of hydrogen ions
    (D) it cannot donate or accept electrons
23. PVC is obtained by the polymerisation of –
    (A) Propene (B) Vinyl Chloride
    (C) Styrene (D) Acetylene
24. Which of the following metals can displace zinc from a solution of zinc sulphate?
    (A) Lead (B) Magnetism
    (C) Iron (D) Mercury
25. Water gas is prepared by passing –
    (A) Steam over white hot coke
    (B) Air over red hot coke
    (C) Methane over red hot coke
    (D) Steam over heated sulphur

26. Which of the following metals react with nitrogen to form nitride?
    (A) Sodium (B) Potassium
    (C) Magnesium (D) None of these
27. One micron is equal to –
    (A) 1/10th of mm (B) 1/100th of mm
    (C) 1/1000th of mm (D) 1/10000th of mm
28. Hydrometer is an instrument –
    (A) For measuring sound in water
    (B) To detect the presence of hydrogen in the atmosphere
    (C) For measuring the specific gravity of liquids
    (D) To detect the changes in atmospheric humidity
29. Which thermometer is used to indicate the lowest temperature ?
    (A) Clinical thermometer
    (B) Gas thermometer
    (C) Alcohol thermometer
    (D) Resistance thermometer
30. A chronometer measures –
    (A) Sound waves (B) Time
    (C) Water waves (D) Colour contrast
31. One fathom is equal to –
    (A) 6 metres (B) 6 feet
    (C) 60 feet (D) 60 cms
32. What is the unit for measuring the pitch or frequency of sound ?
    (A) Coulomb (B) Hum (C) Cycles (D) Decible
33. The fastest acting enzyme in the biological kingdom is –
    (A) Lipase (B) Amylase
    (C) Carboxypeptidase (D) Carbonic anhydrase
34. German biochemist, Emil Fischer –
    (A) Gave Operon model
    (B) Compared the fit between enzyme substrate to lock and key

(C) Proposed the theory of competitive inhibition
(D) Proposed the feed back inhibition theory

35. Enzymes generally have –
(A) same pH and temperature optima
(B) same pH but different temperature optima
(C) different pH but same temperature optima
(D) different pH and temperature optima

36. Coenzyme is –
(A) always protein (B) often metal
(C) always inorganic compound (D) often a vitamin

37. J. B. Sumner isolated first enzyme from Jackbeans as –
(A) amylase (B) trypsin (C) urease (D) rennin

38. Enzymes are absent in –
(A) fungi (B) bacteria (C) viruses (D) algae

39. The enzymes sucrase acts on –
(A) sucrose only
(B) sucrose and starch
(C) all disaccharides
(D) any organic monomer

40. Vitamins are –
(A) inorganic substances which cannot be synthesized by animals
(B) inorganic by animals
(C) organic substances that cannot be generally synthesized by animals
(D) organic substances which can be commonly synthesized by animals

41. Richest source of vitamin B1 (thiamine) is –
(A) cod liver oil (B) curd
(C) whole bread meal (D) eggs

42. One of these vitamins is called erythrocyte maturation factor –
(A) A (B) C (C) K (D) B_{12}

43. The vitamin which can be synthesized in the gut by bacterial action is –
    (A) vitamin B_1 (B) vitmain B_2
    (C) vitmain B_{12} (D) All above
44. Deficiency of vitamin D causes –
    (A) rickets (B) beri-beri
    (C) scurvy (D) night blindness
45. Which one of the following is the best source of vitamin A?
    (A) Apple (B) Carrot
    (C) Amla (D) Raddish
46. Ascorbic acid is –
    (A) Vitamin K (B) Vitamin B
    (C) Vitamin A (D)Vitamin C
47. Sterility in some animals is caused due to deficiency of –
    (A) vitamin B (B) vitamin D
    (C) vitamin B_{12} (D) vitamin E
48. At the boiling temperature an enzyme becomes –
    (A) killed (B) denatured
    (C) unaffected (D) inactivated
49. Starch hydrolysing enzyme is –
    (A) sucrose (B) invertase
    (C) amylase (D) cellulose
50. The enzyme used for the formation of RNA on DNA is –
    (A) DNA polymerase (B) endonuclease
    (C) topoisomerase (D) RNA polymerase

Answer Keys (General Science Objective Type) - V

1. (b)	2. (c)	3. (d)	4. (d)	5. (c)	6. (c)	7. (c)	8. (b)	9. (b)	10. (c)
11. (c)	12. (b)	13. (d)	14. (a)	15. (d)	16. (a)	17. (c)	18. (d)	19. (c)	20. (d)
21. (c)	22. (c)	23. (b)	24. (b)	25. (a)	26. (c)	27. (c)	28. (c)	29. (c)	30. (b)
31. (b)	32. (d)	33. (d)	34. (b)	35. (c)	36. (d)	37. (c)	38. (c)	39. (a)	40. (c)
41. (c)	42. (d)	43. (d)	44. (a)	45. (b)	46. (d)	47. (d)	48. (b)	49. (c)	50. (d)

GENERAL SCIENCE
OBJECTIVE TYPE – VI

1. The term 'enzyme' was coined by –
   (A) de Duve (B) Kunhe
   (C) C. Darwin (D) de Candolle
2. Which set of conditions represents easiest way to liquify gas ?
   (A) Low temperature, high pressure
   (B) High temperature, low pressure
   (C) Low temperature, low pressure
   (D) High temperautre, high pressure
3. Artificial rain is caused by spraying small pellets of which of the following over clouds ?
   (A) Ice (B) Dry Ice (C) NaOH (D) NH4Cl
4. Galvanisation is deposition of –
   (A) Zinc of iron (B) aluminium on Iron
   (C) tin on iron (D) copper on iron
5. Equal volumes of different gases at any definite temperature and pressure have –
   (A) Equal weights (B) equal masses
   (C) equal density (D) equal no. of molecules
6. Milk is an example of –
   (A) suspension (B) true solution
   (C) emulsion (D) gel
7. Which of the following modes of expressing concentration is independent of temperature –
   (A) Molarity (B) Molality
   (C) Normality (D) Formality
8. If temperature is kept constant during the reaction the process is called –
   (A) Adiabatic (B) Isobaric
   (C) Isochoric (D) Isothermal

9. Cholera is caused by –
   (A) vibrio cholera (B) salmonella
   (C) bacterioceoccurs (D) None of these
10. Why do colloidal substance do not deposited below ?
   (A) Specific gravity is less
   (B) It is of very less amount in respect of solvent
   (C) They carry electrical charges
   (D) They are converted to sovent
11. Lightening cause rainfall because –
   (A) It cause combination of oxygen and nitrogen
   (B) Some of the gas molecules become bigger
   (C) It activate H_2O molecule
   (D) Photo-electricity reaction starts
12. Nitrification means –
   (A) Liquifaction of nitrogen
   (B) Convert the atmospheric nitrogen to effective nitrogen compound.
   (C) Production of nitrogen from air
   (D) Conversion of nitrogen to nitric acid.
13. Cell membrane is –
   (A) permeable (B) selectively permeable
   (C) semipermeable (D) impermeable
14. Cholesterol is one kind of –
   (A) Saturated fatty acid (B) unsaturated fatty acid
   (C) Steroid (D) Diglyceride
15. Ethyl-alcohol is mixed completely with water. How ethyl alcohol is separated from the mixture ?
   (A) by separation flannel (B) by evaporation
   (C) partial-distillation (D) evaporation of water
16. White vitriol is –
   (A) $FeSO_4, 7H_2O$ (B) $ZnSO_4, 7H_2O$
   (C) $MgSO_4, 7H_2O$ (D) $CuSO_4, 7H_2O$

17. C.T. Scanning uses –
    (A) Ultrasound waves (B) Gamma rays
    (C) X-rays (D) None of the above
18. Chemically 'speropolenin' is a / an –
    (A) co-polymer of carotinoid and fatty acid
    (B) Carbohydrate
    (C) Propene
    (D) Lactic acid
19. A mixture of salt and sand can be separated by –
    (A) Sublimation (B) Dissolving water
    (C) Gravity separation (D) Dry distillation
20. Saponification involves the hydrolysis of fats and oils by –
    (A) water (B) washing soda
    (C) stearic acid (D) caustic soda
21. Photo-synthesis is a/ an –
    (A) exothermic process (B) endothermic process
    (C) a neutral process (D) a thermostatic process
22. Which of the following is a physical change ?
    (A) oxidation (B) reduction
    (C) sublimation (D) decomposition
23. Which of the following is not a chemical reaction ?
    (A) Burning of paper
    (B) Digestion of food
    (C) Conversion of water into steam
    (D) Burning of coal
24. Which of the following is a chemical change ?
    (A) Rusting of iron (B) Tempering of iron
    (C) Melting of iron (D) Bending of iron
25. Photoelectric effect was discovered by –
    (A) Einstein (B) Hertz (C) Bohr (D) Plank
26. The penetrating power is maximum in –
    (A) a-rays (B) b-rays
    (C) g-rays (D) cathode rays

27. Which of the following rays carry no charge ?
    (A) Cathode rays (B) X-rays
    (C) a-rays (D) Positive rays
28. An atom bomb is based on the principle of –
    (A) Chemical reaction
    (B) Chain reaction in the form of fission
    (C) Nuclear fusion
    (D) None
29. A plant cell differs from an animal cell in the absence of –
    (A) mitochondria
    (B) centrioles
    (C) ribosomes
    (D) endoplasmic reticulum
30. A Vitamin, the absence of which leads to sterility in both males and females is –
    (A) Vitamin K
    (B) Vitamin C or ascorbic acid
    (C) Vitamin D or calciferol
    (D) Vitamin E or tocopherol
31. Anaemia is caused in man due to the deficiency of –
    (A) folic acid (B) vitamin A
    (C) vitamin B_{12} (D) none
32. Antibiotics are drugs used for the cure of –
    (A) Bacterial diseases (B) Cancer
    (C) Malaria (D) Nervous disorders
33. Blood does not coagulate inside the body due to the presence of –
    (A) haemoglobin (B) heparin
    (C) fibrin (D) plasma
34. Blood pressure is dependent on –
    (A) systolic force (B) cardiac output
    (C) peripheral resistance (D) all of these

35. Which of the following substances undergoes chemical change on heating ?
    (A) Sodium chloride (B) Silica
    (C) Lead nitrate (D) Platinum wire
36. Vulcanisation is a –
    (A) Science of studying volcanoes
    (B) method of degrade polymers
    (C) Process connected with the making of voltaic cells
    (D) Process of hardening of rubber by heating it with sulphur.
37. Water gas is prepared by passing –
    (A) steam over white-hot coke
    (B) air over red-hot coke
    (C) methane over red-hot coke
    (D) steam over heated sulphur
38. Which of the following gases is obtained by the reaction of water with calcium carbide –
    (A) Acetylene (B) Ethylene
    (C) Methane (D) Water gas
39. Cooking oil can be converted into vegetable ghee by the process of –
    (A) oxidation (B) hydrogenation
    (C) distillation (D) crystalisation
40. The contact process is involved in the manufacture of –
    (A) nitric acid (B) Sulphuric acid
    (C) Ammonia (D) Caustic soda
41. Polythene is industrially prepared by the polymerisation of –
    (A) methane (B) styrene
    (C) acetylene (D) ethylene
42. Combustion is a process by which –
    (A) heat is produced
    (B) light is produced

(C) heat and light both are produced
(D) None of these

43. PVC is obtained by the polymerisation of –
(A) propane (B) vinyl chloride
(C) styrene (D) Acetylene

44. The process by which an organic compound breaks down into simpler compounds on heating to high temperature is known as –
(A) Aromatisation (B) Polymerisation
(C) Pyrolysis (D) Reduction

45. The hydrogenation of the vegetables oils takes place in the presence of finely divided –
(A) aluminium (B) charcoal
(C) silica (D) nickel

46. In which of the following processes light energy is converted into chemical energy?
(A) Respiration (B) Fermentation
(C) Photosynthesis (D) Photorespiration

47. Rusting of iron involves –
(A) oxidation (B) reduction
(C) decomposition (D) displacement

48. LPG contains –
(A) Butane and isobutene
(B) Butane and Propane
(C) Isobutane and Propane
(D) Butane, Isobutane and Propane

49. The hormone which stimulates milk production in mammal is known as –
(A) Glucagon (B) prolactin
(C) progesterone (D) oestrogen

50. Antidiuretic hormone (ADH) –
(A) inhibits the secretion of growth hormone by the pituitary

(B) inhibits the rate of heart beat

(C) causes the muscular walls of the arterioles to contract that simulates the reabsorption of water from the kidney tubules.

(D) causes an increase in the blood pressure

Answer Keys (General Science Objective Type - VI)

1. (b)	2. (a)	3. (c)	4. (a)	5. (d)	6. (c)	7. (b)	8. (d)	9. (a)	10. (a)
11. (c)	12. (b)	13. (c)	14. (a)	15. (c)	16. (b)	17. (a).	18. (a)	19. (b)	20. (d)
21. (b)	22. (c)	23. (c)	24. (a)	25. (a)	26. (a)	27. (b)	28. (b)	29. (b)	30. (d)
31. (a)	32. (a)	33. (b)	34. (d)	35. (c)	36. (d)	37. (a)	38. (a)	39. (b)	40. (b)
41. (d)	42. (c)	43. (b)	44. (c)	45. (d)	46. (c)	47. (a)	48. (b)	49. (b)	50. (c)

GENERAL KNOWLEDGE ON AGRICULTURE - I

1. **The Himalayan rivers are-**
   (a) monsoon fed (b) snow fed
   (c) ephimeral (d) seasonal
2. **From which of the following states does the river Cauvery originate ?**
   (a) Andhra Pradesh (b) Tamil Nadu
   (c) Karnataka (d) Madhya Pradesh
3. **The river basin which is called "Ruhr of India" is-**
   (a) Damodar (b) Hooghly
   (c) Godavari (d) Swarnarekha
4. **The Siachin glacier is in-**
   (a) Jammu (b) Aksaichin
   (c) Ladakh (d) Garhwal
5. **In which state Maikal range is situated ?**
   (a) Jammu & Kashmir (b) Rajasthan
   (c) Gujarat (d) Madhya Pradesh
6. **Maharashtra plataeu is made up of-**
   (a) alluvial soil (b) coral reef
   (c) sand stone (d) lava
7. **The widest gap across the western Ghats is-**
   (a) Bhor Ghat (b) Thal Ghat
   (c) Khandwa Gap (d) Palghat
8. **On which of the following states retreating monsoon has more effect?**
   (a) Orissa (b) West Bengal
   (c) Punjab (d) Tamil Nadu
9. **The state in India having dry season for only three or four months every year is-**
   (a) West Bengal (b) Mizoram
   (c) Himachal Pradesh (d) Kerala

10. North-East monsoon brings maximum rainfall to-
(a) Mumbai (b) Delhi
(c) Kolkata (d) Chennai

11. The driest place in India is-
(a) Leh (b) Banner
(c) Jaisalmer (d) Bikaner

12. Which of the following states touches the boundary four nations ?
(a) Assam (b) Uttar Pradesh
(c) Nagaland (d) Jammu and Kashmir

13. Which of the following lattitudes passes through India?
(a) Equator (b) Tropic of cancer
(c) Tropic of Capricorn (d) Arctic circle

14. The India-Pakistan boundary was settled by-
(a) McMohan line (b) Radcliff line
(c) Durand line (d) Maginot line

15. Which of the following Indian states is land locked ?
(a) Andhra Pradesh (b) Bihar
(c) West Bengal (d) Kerala

16. The southernmost tip of India is-
(a) Lakshadweep (b) Andaman islands
(c) Kanyakumari (d) Rameshwaram

17. Which is the important red soil area in India ?
(a) Indo-Gangetic plain (b) Southern Peninsula
(c) Gujarat and Maharashtra (d) Jammu & Kashmir

18. Which of the following soils is very hard to cultivate ?
(a) Alluvial (b) Red (c) Black (d) Sandy

19. Which area of Madhya Pradesh is known for white tigers?
(a) Rohilkhand (b) Baghel khand
(c) Malwa (d) None of these

20. Telgu Ganga Project is planned to supply drinking water to-
(a) Hyderabad (b) Coimbatore
(c) Chennai (d) Bangalore

21. Thein Dam is being constructed on the river-
(a) Satluj (b) Ravi
(c) Beas (d) None of these

22. Dry farming in India is extensively practised in-
(a) Kanara plains (b) Deccan region
(c) Coromandal plains (d) Punjab plains

23. In which state of India wet agriculture is practised?
(a) Tamil Nadu (b) Karnataka
(c) Andhra Pradesh (d) Kerala

24. Jhum cultivation is prevalent in-
(a) Punjab (b) Haryana
(c) Madhya Pradesh (d) Manipur

25. The highest milk producing breed of goat in India is-
(a) Barbari (b) Beetal
(c) Jamnapari (d) Black Bengal

26. After u. p. which state leads in production of sugarcane ?
(a) Tamil Nadu (b) Maharashtra
(c) Bihar (d) Andhra Pradesh

27. In which of the following states does the maximum density of canals lie?
(a) Tamil Nadu (b) Andhra Pradesh
(c) Bihar (d) U. P.

28. Mathura refinery receives crude oil from-
(a) Cauvery basin (b) Vadodara
(c) Kandla (d) None of these

29. In which of the following states Koyali having a petroleum refinery is located ?
(a) Bihar (b) Assam (c) Gujarat (d) U. P.

30. Lignite is found in India in-
(a) Rourkela (b) Hazaribagh
(c) Raniganj (d) Neyveli

31. The state in south India with a thermal station at Tuticoran is-
(a) Karnataka (b) Kerala
(c) Tamil Nadu (d) Andhra Pradesh

32. At present, India's largest mineral resource is-
(a) copper (b) coal
(c) iron-ore (d) none of these

33. The only place where tin is found in India is-
(a) Hazaribagh (b) Rewa (c) Salem (d) Quilon

34. India has the maximum amount of foreign trade with-
(a) U. S. A. (b) China
(c) Japan (d) Germany

35. Nhava Sheva, one of India's major ports is in-
(a) Gujarat (b) West Bengal
(c) Maharashtra (d) Andhra Pradesh

36. Which of the following ports does not have a natural harbour?
(a) Cochin (b) Mumbai
(c) Chennai (d) Vishakhapatnam

37. The first Electric train in India (Deccan Queen) was run between-
(a) Mumbai and Surat
(b) Howrah and Delhi
(c) Kalyan and Pune
(d) New Delhi and Chennai

38. Gitanjali Express runs between-
(a) Howrah-Puri (b) Howrah-Jammu
(c) Howrah-Mumbai (d) Howrah-Jaipur

39. The first shore-based steel plant in India is in-
(a) Haldia (b) Kandla
(c) Paradeep (d) Vishakhapatnam

40. The place 'Avadi' which is known for the manufacture of Vijay anta tanks for the Indian Army is located in-
(a) Orissa (b) Karnataka
(c) Andhra Pradesh (d) Tamil Nadu

41. In which industry is sal wood used mostly?
(a) Railway sleepers (b) Matches
(c) Paper (d) None of these

42. The main centre of diamond cutting industry in India is-
(a) Panna (b) Kolar (c) Surat (d) Agra

43. Where is the Hindustan Shipyard located?
(a) Cochin (b) Vishakhapatnam
(c) Mumbai (d) Kolkata

44. The satellite freight city is being developed near which of the following cities?
(a) Kolkata (b) Bangalore
(c) Kandla (d) Gurgaon

45. India's oldest Planetorium is located at-
(a) Kolkata (b) Mumbai
(c) Patna (d) Ahmadabad

46. The first submarine built by India is-
(a) INS Shalki (b) INS Ganga
(c) INS Godavari (d) INS Himgiri

47. Vijayawada is situated on the bank of the river--
(a) Tungabhadra (b) Godavari
(c) Krishna (d) Cauvery

48. NEFA is the old name of-
(a) Karnataka (b) Tamil Nadu
(c) Nagaland (d) Arunachal Pradesh

49. Kaiga in Karnataka is famous for-
(a) nuclear plant (b) ancient fort
(c) fertiliser plant (d) cement plant

50. Palk straits separates India from-
(a) Bangladesh (b) Pakistan
(c) Sri Lanka (d) Maldives

Answer Keys

1. (b)	2. (c)	3. (a)	4. (c)	5. (d)	6. (d)	7. (d)	8. (d)	9. (d)	10. (d)
11. (c)	12. (d)	13. (b)	14. (b)	15. (b)	16. (c)	17. (d).	18. (b)	19. (b)	20. (c)
21. (b)	22. (b)	23. (d)	24. (d)	25. (b)	26. (b)	27. (d)	28. (c)	29. (c)	30. (d)
31. (c)	32. (b)	33. (b)	34. (a)	35. (c)	36. (c)	37. (c)	38. (c)	39. (d)	40. (d)
41. (a)	42. (c)	43. (b)	44. (d)	45. (a)	46. (a)	47. (c)	48. (d)	49. (a)	50. (c)

childhood obesity as a major health issue. According to the CDC, in the United States, as of 2024, "the percentage of children and adolescents affected by obesity has more than tripled since the 1970s. In 2017 about one in five school-aged children was affected by obesity (20.7% of all 6–11 year olds, 21.2% of all 12–19 year olds). The prevalence of obesity among 2–19 year old children was 25.6% for Hispanic children, 24.2% for Black children, 16.1% for White children, and 8.7% for Asian children. Many factors contribute to childhood obesity, including:

- Metabolism—how your body changes food and oxygen into energy it can use.
- Eating and physical activity behaviors.
- Community and neighborhood design and safety.
- Short sleep duration.
- Negative childhood events."[26]

Once again, as part of adolescent physical development, counselors and teachers should be aware of these obesity statistics and possible causes. Some middle school adolescents suffering from obesity and consequent bullying issues may need counselor intervention.

Conclusion

Life for a young middle school student is all about change. In some ways, adapting to a new school environment, new teachers, new classmates, and a new schedule may transition quite smoothly. At the same time, these young adolescents are experiencing significant cognitive developmental growth that will increase their abstract thinking and abilities to synthesize and analyze—the essence of critical thinking. As their executive functioning develops, so will their abilities to inhibit distractions, plan, organize, manage time, and task monitoring. With these changes, hopefully, middle school students are open to future possibilities as they explore possible career paths and college exploration.

At the same time, adolescents are going through biological changes and growth spurts. Puberty brings change to both boys and girls, changes that continue throughout adolescence. This period is also a time of psychosocial development as young adolescents are establishing their self-concept, self-esteem, and identity formation. Middle school students should learn how to handle their emotions and behaviors by gaining self-control. Counselors and teachers involved must help their students to understand the dynamics of early adolescent development, including regulating their emotions and behaviors. Finally, emotional self-regulation and the ability to express oneself will become increasingly more important as they grow. We strongly recommend that counselors have a strong, proactive role in the lives of middle school students as they engage in this all-important stage of their K–12 journeys.

Notes

1 Curry, Jennifer R. and Amy Milsom. *Career and College Readiness Counseling in P-12 Schools,* 2nd edition. New York: Springer Publishing Co., 2017, p. 212.
2 Crow, Lester and Alice Crow. *Adolescent Development and Adjustment,* 2nd edition. New York: McGraw-Hill Book Co., 1965.
3 Rhodes, S. et al. "Developmental Changes in the Engagement of Episodic Retrieval Processes and their Relationship with Working Memory during Middle Childhood." *British Journal of Developmental Psychology.* https://doi.org/1111/j.2044-835x.2010.02014
4 Rhodes, S. et al. "Developmental Changes in the Engagement of Episodic Retrieval Processes and their Relationship with Working Memory during Middle Childhood." *British Journal of Developmental Psychology.* https://doi.org/1111/j.2044-835x.2010.02014
5 Rhodes, S. et al. "Developmental Changes in the Engagement of Episodic Retrieval Processes and their Relationship with Working Memory during Middle Childhood." *British Journal of Developmental Psychology.* https://doi.org/1111/j.2044-835x.2010.02014
6 Wigfield, Alan, et al. "Early Adolescents' Development Across the Middle School Years: Implications for School Counselors." *Professional School Counseling,* December 2005, Vol. 9, No. 2, Special Issue: Focus On Middle School Counseling, p. 113.
7 Wigfield, Alan, et al. "Early Adolescents' Development Across the Middle School Years: Implications for School Counselors." *Professional School Counseling,* December 2005, Vol. 9, No. 2, Special Issue: Focus On Middle School Counseling, p. 113.
8 Wigfield, Alan, et al. "Early Adolescents' Development Across the Middle School Years: Implications for School Counselors." *Professional School Counseling,* December 2005, Vol. 9, No. 2, Special Issue: Focus On Middle School Counseling, p. 113.
9 Wigfield, Alan, et al. "Early Adolescents' Development Across the Middle School Years: Implications for School Counselors." *Professional School Counseling,* December 2005, Vol. 9, No. 2, Special Issue: Focus On Middle School Counseling, p. 113.

10 Wigfield, Alan, et al. "Early Adolescents' Development Across the Middle School Years: Implications for School Counselors." *Professional School Counseling*, December 2005, Vol. 9, No. 2, Special Issue: Focus On Middle School Counseling, p. 113.
11 Wigfield, Alan, et al. "Early Adolescents' Development Across the Middle School Years: Implications for School Counselors." *Professional School Counseling*, December 2005, Vol. 9, No. 2, Special Issue: Focus On Middle School Counseling, p. 115.
12 Wigfield, Alan, et al. "Early Adolescents' Development Across the Middle School Years: Implications for School Counselors." *Professional School Counseling*, December 2005, Vol. 9, No. 2, Special Issue: Focus On Middle School Counseling, p. 115.
13 Gaertner, Matthew and Katie McClarity. "Performance, Perseverance, and the Full College Readiness." *Educational Measurement Issues and Practice*, Summer 2015, Vol. 34, No. 2, pp. 20–33.
14 Gaertner, Matthew and Katie McClarity. "Performance, Perseverance, and the Full College Readiness." *Educational Measurement Issues and Practice*, Summer 2015, Vol. 34, No. 2, pp. 20–33.
15 Erikson, E. H. *Childhood and Society*, 2nd edition. New York: W.W. Norton & Co., 1993.
16 Marcia, James, et al. *Ego Identity: A Handbook for Psychosocial Research*, 2nd edition. New York: Springer-Verlag, 1993.
17 Marcia, James, et al. *Ego Identity: A Handbook for Psychosocial Research*, 2nd edition. New York: Springer-Verlag, 1993.
18 Curry, Jennifer R. and Amy Milsom. *Career and College Readiness Counseling in P-12 Schools*, 2nd edition. New York: Springer Publishing Co., 2017, p. 231.
19 Wigfield, Alan, et al. "Early Adolescents' Development Across the Middle School Years: Implications for School Counselors." *Professional School Counseling*, December 2005, Vol. 9, No. 2, Special Issue: Focus On Middle School Counseling, p. 113.
20 Wigfield, Alan, et al. "Early Adolescents' Development Across the Middle School Years: Implications for School Counselors." *Professional School Counseling*, December 2005, Vol. 9, No. 2, Special Issue: Focus On Middle School Counseling, p. 113.
21 Lemberger-Truelove, Matthew et al. "Growth in Middle School Student's Curiosity, Executive Functioning and Academic Achievement." *Professional School Counseling*, May 2021, Vol. 24, No. 1b, Special Issue: Educator–Counselor: A Nondual Identity for School Counselors, p. 1.
22 Lemberger-Truelove, Matthew et al. "Growth in Middle School Student's Curiosity, Executive Functioning and Academic Achievement." *Professional School Counseling*, May 2021, Vol. 24, No. 1b, Special Issue: Educator–Counselor: A Nondual Identity for School Counselors, p. 1.
23 American School Counselor Association National Model (ASCA, 2025). https://schoolcounselor.org/About-School-Counseling/ASCA-National-Model-for-School-Counseling-Programs
24 Curry, Jennifer R. and Amy Milsom. *Career and College Readiness Counseling in P-12 Schools*, 2nd edition. New York: Springer Publishing Co., 2017, p. 212.
25 "Caring for Your School-Age Child: Ages 5–12." *American Academy of Pediatrics, Healthy Children*. www.Healthychildren.org
26 Obesity | Healthy Schools | CDC. Aug 2022. https://archive.cdc.gov/www_cdc_gov/healthyschools/obesity/index.htm#

3

The Social and Emotional Needs of Middle School Students

How to Succeed in Overall Functioning in Mental and Emotional Areas

As students walk into middle school, not only are they entering a more demanding academic stage of their young lives, but they are also experiencing a higher level of social and emotional learning (SEL). In these formative years, the role of the counselor may have a great impact on students' developing sense of themselves. School counselors who have professional experience in child and adolescent development play an important role in the education, prevention, and intervention activities that may be integrated into the student experience. Unfortunately, in many schools, the role of the school counselor is "encumbered with tasks that distance them from formal interventions intended to improve circumstances for students and other members of the school community."[1] Consequently, such routine tasks make it difficult for counselors to work directly with students, clearly interfering with the positive effect of counselor intervention on the social and emotional learning of middle school students. In addition, we refer to other sources, both the Wigfield et al. and the Lemberger-Truelove et al. studies, that advocate both teacher and counselor intervention. These studies are also referenced in

DOI: 10.4324/9781003597292-4

Chapter 4 confirming the effectiveness of teacher/counselor intervention in the academic achievement of middle school students.

Social and Emotional Learning

Social and emotional learning "is the process through which children and adults develop the skills, attitudes and values necessary to acquire social and emotional competence."[2] In the classroom, this may be developed by providing instructional activities that foster students' ability to understand, manage, and express their feelings and to engage in rewarding interactions with others. As Tayor and Larson note, Howard Gardner "addresses the skills and aptitudes of social and emotional learning as inter- and intrapersonal intelligence which starts from and builds on students' innate ability to understand themselves and others, stemming from their own abilities, emotions, needs and points of reference and others' feelings and perspectives."[3] Middle school students are capable of developing their social and emotional learning as they build on their own abilities to understand themselves and others.

SEL provides many implications for counselors, psychologists, social workers, and therapists. In 1997, Albert Bandura presented a framework for social and emotional learning that includes self-efficacy playing a pivotal role.[4] Bandura's self-efficacy theory stresses that ability perception appears to have a major impact on choice behavior and defines self-efficacy as the expectation to successfully perform a behavior required for a particular task. In the life of a middle schooler, this self-perception may have a powerful impact on their view of themselves not only academically, but also socially and emotionally. Helping adolescents achieve an enhanced self-perception that influences choice behavior and defines their belief that they can successfully perform a task is a necessary component of the middle school education process.

According to Jones et al., Bandura states that self-efficacy is a component of "social-cognitive theory that speaks to the foundation of self-efficacy beliefs, their structure and functional properties, their varied effects, the manner in which they work, and the progression by which one enlists such beliefs for personal and social change."[5] In making choices, children are clearly influenced by their self-perceptions and sense of competence. Jones et al. further refer to Bandura's development of the theory and foundations of self-efficacy that posits that people's beliefs regarding their capabilities vary across activities and situational circumstances, rather than manifesting uniformly across tasks and contexts. For example, in the life of young middle school students, their skills and achievements in given classes may vary widely. Students know where they feel more accomplished and where they may need more assistance. This variance applies to activities and sports as well.

Jones et al. note that Bandura suggests that there are four ways to establish a foundation for self-efficacy. The first way is grounded in how a person reacts to experiencing difficulty with a task, especially when they have experienced "first-mastery". How people respond is critical in developing resiliency, a key component of self-efficacy. Bandura's second way is rooted in social modeling, also known as *subconscious learning,* which plays an influential role in developing self-efficacy through the acquisition of attitudes, values, and styles of behavior. Third, Bandura lists social persuasion as a major influence. A person who is persuaded to believe in him/herself increases perseverance when that person encounters struggles. A more positive self-assessment of their capabilities results when individuals learn to believe in themselves, which increases their perseverance when people face struggles. Bandura's (1997, 2001) fourth way suggests that self-efficacy beliefs contribute to self-development. "Such beliefs impact the array of options people may consider and the choices they make when deciding on setting the course of their life paths."[6] As noted by Jones et al., Bandura's theory suggests much work for counselors. In regard to some SEL interventions, counselors should focus on how children interpret social cues and respond to social challenges. Jones et al. have further advised that for social learning to occur, children and adolescents must exchange knowledge in an interactive environment. When middle school children see someone experiencing success through perseverance, that observation may increase their aspirations and beliefs in their own potentials. Referencing Bandura, Jones et al. further state Bandura's argument that social persuasion, being persuaded to believe in oneself, could be a mode of influence, helping to increase the potential for positive social outcomes.[7]

Furthermore, the Lemberger-Truelove et al. study[8] researching the impact of counselor assistance found that school counselor intervention to be successful in the academic achievement and social and emotional learning of middle school students. This study compared a cluster-randomized design at the classroom level composed of two ethnically diverse and economically challenged samples of 109 urban middle school students divided between a treatment group and a delayed treatment group. Using the Advocating Student-within-Environment (ASE) school counseling approach, this study focused primarily on the SEL and MBI (Mindfulness-Based Interventions) practice aspects of the ASE approach. The results of this study exhibited greater positive changes in stress tolerance, social curiosity, executive functioning, and academic achievement. "Most importantly, results from the current study demonstrates the developmental and learning potential of all students, particularly minoritized and economically challenged students who are often erroneously considered from a deficit position."[9] Additionally, the results of this study demonstrate that counselor intervention that included learning strategies, self-management skills, and social skills were delivered in classrooms as well as through consultation activities with teachers. These activities are specific recommendations offered in the ASCA Mindsets and Behaviors as presented by Curry and Milsom.[10]

Wigfield et al. also stressed that the former National Middle School Association (now, the Association of Middle Level Education) recognizes the complex social and educational transition that is occurring in the middle school years and encourages counselors to be proactive in this transition.

> Specific recommendations include: (a) counselors working collaboratively with teachers and counselors in elementary schools, particularly to acknowledge curricular differences across schools; (b) counselors educating parents about issues related to the middle school transition; (c) the

> importance of counseling about the transition at both the elementary and middle school levels; and (d) counselors providing activities, curricula, and services aimed specifically at easing the transition to middle school.[11]

Incorporating these recommendations into middle schools may help to ease the student's transition to their new environment. Furthermore, Wigfield et al. also suggests that cooperation among teachers, counselors, and administrators to ease this transition and further advises that parental involvement is essential in a successful transition. (Chapter 14 of Part III presents a consistent methodology of communicating with parents as students progress through middle school.) Part II of this book explores the many ways that teachers, counselors and administrators may assist students in their transition to middle school, particularly as they learn to develop more advanced study skills and habits. In addition, the advisory/group counseling programs accent the importance of goal-driven tasks for academic achievement and then move into exploring possible careers and colleges. The Wigfield et al. study concludes by recommending that counselors should also assist students in helping middle school students negotiate their many new experiences as well as assisting with personal issues, cultural awareness, and sensitivity advocacy for attention to the diverse needs of all students.[12]

A side note: As noted in our prior book, *Counseling 21st Century Students for Optimal College and Career Readiness, A 9th-12th Grade Curriculum* Second Edition, we as coauthors understand that counselors have always played a critical role in the development of their students. As the world becomes more complex and as students are increasingly exposed to many forces during their developmental years, our counselor roles take on even more importance. Both of us have experienced the roles of counselor, teacher, and administrator; this collective experience has broadened our perspectives of what adolescents need over time for successful growth and development. Our experience with middle school students has only reinforced our long-held belief that helping students involves the work of all—teachers, counselors, administrators, parents, and students alike—as adolescents successfully navigate their ways through middle school. We also recognize the difficulties that many middle school students may experience. At the same time, we recognize that middle school will hopefully prepare students for the next step of their K–12 journeys as they successfully transition to high school.

Moreover, David S. Yeager's *Social and Emotional Learning Programs for Adolescents* presents a strong argument for the need for effective SEL programs. While he acknowledges the more traditional SEL programs, he questions how well they work for adolescents. "First, effective universal social-emotional learning can transform young people's lives for the better. Effective programs can prevent catastrophic outcomes, such as unwanted pregnancy, arrests for violent crime, or dropping out of high school. They can also encourage greater thriving, including having less stress, better health, and a greater love of learning." In addition, Yeager acknowledges that improving adolescents' interior social and emotional lives can impact other areas of functioning, "because social and emotional life matters so much at this age."[13] How middle school educators address helping adolescents at this critical time is so important in the lives of students. As educators strive to create a positive learning environment that fosters empathy, honesty, and integrity, more positive social and emotional growth will occur. Counselors, teachers, and administrators must also have an awareness of possible negative aspects of poor social and emotional development so that they may readily address problematic issues. Targeted negative behaviors should include bullying, substance prevention among others and must be addressed as these behaviors harm individual students as well as the overall school community.

In describing the needs of young adolescents, Yeager addresses both biological and psychosocial changes and refers to the work of Bradford Brown, a developmental psychologist at the University of Wisconsin. In a report for the National Academy of Sciences, Brown established that adolescents have four developmental tasks:

1. To stand out: to develop an identity and pursue autonomy;
2. To fit in: to find comfortable affiliations and gain acceptance from peers;

3. To measure up: to develop competence and find ways to achieve, and
4. To take hold: to make commitments to particular goals, activities, and beliefs.

Yeager posits that when SEL programs "honor adolescents' desire to achieve these tasks, the program works because it is respecting the kind of person an adolescent wants to become, motivating them to change."[14] At the basis of the four developmental tasks is the adolescents' understanding of who they are and what they may need to feel comfortable as they develop competencies that result in achievements. The fourth developmental task results in the individual's ability to enter into commitments in goal-directed activities.

In effect, Yeager advocates a combined climate and mindsets model. The climate and mindsets model tries

> to motivate young people in terms of the values that matter most to them, and they try to change how young people see the world- their mindsets. Effective programs align the adult-sanctioned healthy choice-not getting pregnant, not getting arrested, etc.- with peer-sanctioned sources of status and respect like freedom, autonomy, or mattering. These programs do this both in how they talk to young people- by offering opportunities for authentic choice and input- and in what they teach- by helping young people envision a desirable future as the kind of person who makes healthy choices.[15]

In effect, this model helps students to develop an approach to life in a positive, healthy manner. According to Yeager, broadening the approach to SEL programs to create positive climates within the school and healthy mindsets that help adolescents cope more successfully with the challenges that they encounter requires significant work on the part of educators. However, the evidence is so encouraging that it demands urgent action in schools across the country.[16]

Mental Health Challenges: The Need for Adolescent Mental Health

At the same time that middle school students are experiencing cognitive and physical growth and changes as well as social and emotional development, an alarming number of adolescents are in need of mental health assistance. Former Surgeon General Vivek Murthy was Surgeon General from 2014 to 2017 and from 2021 to January 2025. Like former 1980s Surgeon General Dr. C. Everett Coop who famously addressed the health dangers associated with smoking, Vivek Murthy in 2021 issued a rare warning calling national attention to adolescent mental health. As teachers, counselors, administrators, students, and parents, this warning is of utmost importance to all.

Two years later, in an interview with the *New York Times* in 2023, Dr. Vivek Murthy ascribed the mental health challenges among young people in part to "hustle culture values." His warning was that Mental Health Challenges were leading to "devastating effects" among young people, including middle school children. His statements came as the suicide rate for young Americans **ages 10–19** jumped by 40% from 2001 to 2019, while emergency room visits from self-harm rose by 88%. In 2023, he spoke with the *New York Times* about the "growing intensity" of his focus. We provide some of these concerns that were discussed in his session with the *New York Times'* Matt Richtel. Murthy has described mental health as "the defining public health crisis of our time."[17]

Murthy thinks mental health allows us to show up for our communities, friends, family, and our lives. "And when the fuel tank is low, it compromises our ability to be who we can be in our best moments and to show up at work or in school." Yet, he views the challenges as getting worse. He noted that hospitalizations have gone up, even for young adolescents. "Something is driving the greater pain and despair." In addition, he notes a number of crises: loneliness and isolation; bullying in person and online.

> The fact that our kids are surrounded by an information environment that is coming at them 24/7 and that often

> stokes fear and anxiety. It's also being driven by the fact that young people, when they think about the future, see the profound threats that we are facing today, like violence and racism and climate change, but they don't see effective solutions.

Murthy believes that young people, especially in "early adolescence," are in a "very sensitive phase of development." He has noted the following as so crucial to understand when he talks to young people who consistently tell him three things about social media: "they say it makes them feel worse about themselves; about their friendships; and they can't "get off it."" Murthy adds that while adolescents are spending so much time online, that time is taking away from their sleep, from exercise, from in-person interaction with people, from schoolwork, and from other activities that might bring them joy.

Finally, Murthy was asked: "What's the connection between Values and Mental Health?" He responded:

> Values are the filter that we use to help us make decisions when we are faced with choices. But our values are also what inform what we drive toward in life. Young people tell me they get caught up in the hustle culture. What they're saying to me was that they felt that they were being asked to chase certain objectives – getting a job with a fancy title, making a lot of money, becoming famous, acquiring power. And not only did many of them say they were exhausted, but they weren't sure that was going to bring them happiness."

Dr. Murthy ended his discussion accordingly: "Many of them feel exhausted...Are we pushing our kids to pursue what's really going to bring them happiness and their fulfillment." It is essential to keep in mind that the Surgeon General's report is not just based on statistics. Murthy has come to these conclusions after a lengthy listening tour speaking with students, health care workers, and community groups across the country. His service is remarkable in defining what is not clearly a hazardous, risky behavior such as smoking, but rather a mental health crisis that is afflicting our nation.[18]

*Please note that excessive use of social media and doomscrolling is addressed in Chapter 5.

Recommendations for the Role and Responsibilities of School-Based Mental Health Counselors with Middle School Children

The American Mental Health Counselors Association (AMHCA) was created to recognize and support non-school counselors in community and agency settings in 1979. Licensed mental health counselors serve in a range of clinical practices that may include psychoeducational techniques aimed at the prevention of mental and emotional disorders, consultations to individuals, couples, and "diagnosis and treatment of mental and emotional disorders." Because school- aged children are more likely to receive mental health services provided in schools than at community agencies, Professors David D. Christian, and Cian L. Brown have recommended the development of a School Based Mental Health Counselor (SBMHC).[19] In addition, they indicate the creation of training standards and credentialing for SBMHCs due to their ever-increasing role in delivering mental health services to school-aged children. As coauthors, we have both experienced the many evolving roles and tasks of the school counselor. Clearly, this call for mental health counselors is a necessary addition to good educational practices.

Dr. Fitzpatrick developed a 60 credit Mental Health Master's Degree Counseling Program at Manhattan University in New York City (2006). This program was accepted by the New York State Office of Professions in late 2006. From its beginning, the program has been very active in working in school settings where the need for mental health counselors is essential. Like psychologists and social workers, Mental Health Counselors must file in the New York State Office of Professions. Once they graduate from their program, they must work with a "limited" license for approximately 3000 hours of postgraduate work before they can practice full time. It is important to understand how much Mental Health Counselors must learn and experience first-hand in the schools where the need for clinical services to engage students in their development is so clearly evident. Often schools and nonprofit organizations accept the graduates with their limited status. Most Mental Health Counselors join the American Mental Health Counselors Association (AMHCA). The association is a nonprofit organization for clinical mental health counselors dedicated to advancing the profession.

As noted by John D. Preston et al. in *Child and Adolescent Clinical Psychopharmacology Made Simple 3rd Edition*, most children and teenagers suffering from psychological problems do not require medication treatment. Rather, they may need psychosocial interventions that often involve the family as well.[20] Licensed mental health counselors are able to address this growing need in schools. Dr. Fitzpatrick, as a licensed Clinical Psychologist and a licensed Mental Health Counselor, has been able to advocate for the recognition of having Mental Health Counselors working in school settings. Her evaluation of Christian and Brown's *Recommendations for the Role and Responsibilities of School–Based Mental Health Counselors*[21] understands the authors' recommendations from 2018. Having licensed mental health professionals serve in middle and high schools where they can address mental health issues that interfere with the academic, career, and personal/social success of all students is truly needed.

At the same time, mental health counselors should know the limitations of their training and should refer students and families to appropriate clinical mental health counselors when necessary. These counselors should be licensed by their state to provide mental health services. Licensed mental health counselors are also employed in community mental health centers, hospitals, and nonprofit and for-profit agencies and organizations covered by managed care and health care providers. According to Becker et al. (2015),[22] certified mental health counselors are providing services in schools with growing frequency. As Christian and Brown have noted, many school-aged children are experiencing increased issues that must be addressed by mental health professionals. As many of these children are not receiving adequate care, school-based mental health counselors should be used to address these needs. Furthermore, Christian and Brown recommend that school-based mental health counselors and school counselors partner in conducting research to evaluate school-based services to improve the school's educational policy.

An added note: Many private schools have a licensed psychologist on staff either full-time or part-time to address the mental health issues of the students. Usually, the role of the on-staff psychologist is to evaluate and assess a student's need and make necessary referrals.

A Review of Evidence-Based Practices and Third Wave Therapies Treatment Modalities

Counselors should be sure to review this section as it includes various types of therapies that were known as basically behaviorism and others that are beginning to engage in approaches needed for clients. Judith Scott, Ph.D. and Christin M. Jungers, Ph.D. in their book *Practicum and Internship*[23] have developed evidence-based practices (known as the first generation) that also include traditional behavior as well as what is now called third wave therapies. The second generation included the start of cognitive behavior therapy (CBT). During the 1990s, cognitive and behavioral approaches merged to some extent, leading to the current development of CBT—thoughts, feelings, and behaviors. A third wave of CBT has what is known as an existential component that assumes that suffering is a basic characteristic of human life. More recently, CBT continues to evolve with the "third wave" that has developed Dialectical Behavior Therapy (DBT). CBT is widely used evidence-based Psychotherapy that aims to help individuals identify, challenge, and change negative thought patterns and behaviors that contribute to mental health. CBT also focuses on cognitive restructuring which can be used to challenge negative thoughts. Mindfulness-based cognitive therapy (MBCT) is yet another integration of CBT, particularly as therapy for depression.

Yet another therapy that we hope will be useful to counselors is "Acceptance and Commitment Therapy" (ACT). The theme for this approach includes that a person can take action without first changing or eliminating feelings but rather an effective approach that is to accept and change. Drs. Jungers and Scott have also developed a detailed understanding of the development of Dialectical Behavior Therapy, and it should be on the list of therapies to consider.[24] In summary, the development of cognitive skills for counselors-in-training should include reviewing resources focused on goal setting, treatment planning, and theory-based approaches to treatment. This approach provides enormous implications for counseling with a focus on the wellness model that has more recently been used by counselors, teachers, and administrators in helping middle schoolers achieve and maintain mental and emotional wellness.

Conclusion

Middle school students are in a crucial developmental stage of their young lives. How can we as educators help them to succeed in overall functioning in mental and emotional areas? Clearly, this is a formative period where they are laying the foundation of what Bandura referred to as the time in which adolescents develop a belief in their own self-efficacy. Furthermore, the Lemberger-Truelove et al. study demonstrated a positive outcome in early counselor intervention. Similarly, the Wigfield et al. study also promotes active counselor involvement with students and parents alike in the transition from elementary school to middle school. Yeager also advocates for a new model for social and emotional learning environment that is based upon a positive climate and healthy mindsets during these highly formative middle school years.

At the same time, some of the middle school students are experiencing anxiety, depression, self-harm, and suicidal thoughts. Former Surgeon General Vivek Murthy brings particular attention to this subject. Issues presented in Murthy's Letter call for increased school counselors and mental health counselors. As noted above, the AMHCA as well as the Christian and Brown evaluations recognize the increased need for mental health counselors (non-school counselors) in middle schools.

Again, we recommend that educators—teachers, counselors, and administrators—all work together to promote a more positive growth period for children in middle schools, including thoughts, feelings, and behaviors as part of the growth process. It is essential also to include parents in the overall experience of middle school child development. Part II and Chapters 14 (Working with the Families of Middle School Children) and 15 (The Move to High School: Making Transitioning to 9th Grade Successful) of this book specifically include all—educators, students, and parents—as we address multiple issues from goal development and enhanced study skills to early career and college exploration and transition to high school. Together, we can help middle school students thrive in this all-important stage of their lives.

Notes

1 Lemberger-Truelove, Matthew et al. "Growth in Middle School Student's Curiosity, Executive Functioning and Academic Achievement." *Professional School Counseling*, May 2021, Vol. 24, No. 1b, Special Issue: Educator–Counselor: A Nondual Identity for School Counselors, p. 1.
2 Taylor, Howard E. and Susan Larson. "Social and Emotional Learning in Middle School." *The Clearing House*, Jul–Aug., 1999, Vol. 72, No. 6, p. 331.
3 Taylor, Howard E. and Susan Larson. "Social and Emotional Learning in Middle School." *The Clearing House*, Jul–Aug., 1999, Vol. 72, No. 6, p. 331.
4 Bandura, Albert. *Social Foundation of Thought and Action: A Social Cognitive Theory*. New Jersey: Prentice Hall, Inc, 1986.
5 Jones, Brenda et al. *Child and Adolescent Counseling*. Springer Publishing Co, 2022, pp. 19–20.
6 Jones, Brenda et al. *Child and Adolescent Counseling*. Springer Publishing Co, 2022, pp. 19–20.
7 Jones, Brenda et al. *Child and Adolescent Counseling*. Springer Publishing Co, 2022, pp. 19–20.
8 Lemberger-Truelove, Matthew et al. "Growth in Middle School Student's Curiosity, Executive Functioning and Academic Achievement." *Professional School Counseling*, May 2021, Vol. 24, No. 1b, Special Issue: Educator–Counselor: A Nondual Identity for School Counselors (May 2021), p. 6.
9 Lemberger-Truelove, Matthew et al. "Growth in Middle School Student's Curiosity, Executive Functioning and Academic Achievement." *Professional School Counseling*, May 2021, Vol. 24, No. 1b, Special Issue: Educator–Counselor: A Nondual Identity for School Counselors (May 2021), p. 5.

10 Curry, J. R. and A. Milsom. *Career and College Readiness Counseling in P-12 Schools*, 2nd edition. Spring Publications, 2017, p. 219 and 239.
11 Wigfield, Alan, et al. "Early Adolescents' Development Across the Middle School Years: Implications for School Counselors." *Professional School Counseling*, December 2005, Vol. 9, No. 2, Special Issue: Focus On Middle School Counseling, p. 117.
12 Wigfield, Alan, et al. "Early Adolescents' Development Across the Middle School Years: Implications for School Counselors." *Professional School Counseling*, December 2005, Vol. 9, No. 2, Special Issue: Focus On Middle School Counseling, p. 117.
13 Yeager, David S. "Social and Emotional Learning Programs for Adolescents." *The Future of Children*, Spring 2017, Vol. 27, No. 1, Social and Emotional Learning, Princeton University, pp. 73–94.
14 Yeager, David S. "Social and Emotional Learning Programs for Adolescents." *The Future of Children*, Spring 2017, Vol. 27, No. 1, Social and Emotional Learning, Princeton University, p. 76.
15 Yeager, David S. "Social and Emotional Learning Programs for Adolescents." *The Future of Children*, Spring 2017, Vol. 27, No. 1, Social and Emotional Learning, Princeton University, p. 79.
16 Yeager, David S. "Social and Emotional Learning Programs for Adolescents." *The Future of Children*, Spring 2017, Vol. 27, No. 1, Social and Emotional Learning, Princeton University, pp. 88–89.
17 Richtel, Matt. "The Surgeon General's New Mission: Adolescent Mental Health." *New York Times*, March 21, 2023. https://www.nytimes.com/2023/03/21/health/surgeon-general-adolescents-mental-health.html
18 Richtel, Matt. "The Surgeon General's New Mission: Adolescent Mental Health." *New York Times*, March 21, 2023. https://www.nytimes.com/2023/03/21/health/surgeon-general-adolescents-mental-health.html
19 Christian, David D. and Cian L. Brown. "Recommendations for the Role and Responsibilities of School-Based Mental Health Counselors." School-Based Mental Health Counselors. *Journal of School-Based Counseling Policy and Evaluation*, 2018, pp. 26–39.
20 Preston, John D., et al. *Child and Adolescent Clinical Psychopharmacology Made Simple*, 3rd edition. New Harbinger Publications, Inc, 2015, p. 2.
21 Preston, John D., et al. *Child and Adolescent Clinical Psychopharmacology Made Simple*, 3rd edition. New Harbinger Publications, Inc, 2015, p. 2.
22 Preston, John D., et al. *Child and Adolescent Clinical Psychopharmacology Made Simple*, 3rd edition. New Harbinger Publications, Inc, 2015, p. 30.
23 Junger, Christin M. and Judith Scott. *Practicum and Internship: Textbook and Resource Guide for Counseling and Psychotherapy*, 6th edition. Routledge Press, 2019.
24 Junger, Christin M. and Judith Scott. *Practicum and Internship: Textbook and Resource Guide for Counseling and Psychotherapy*, 6th edition. Routledge Press, 2019.

PART II

Helping Middle School Students Achieve Academic Success and Explore Careers and Colleges

4

Creating an Effective Advisory System or Group Counseling Program

Why an Advisory System or a Group Counseling Program Is Beneficial to Middle School Students

As middle school students enter this new phase of their education, they encounter multiple challenges. The world of elementary school may feel far away to the many middle schoolers who find themselves traveling on a school bus for the first time to a much larger school with an increased student population. Once inside this new middle school, students now encounter changing classes, using lockers, and attempting to keep to their new schedule. At the same time, they are meeting new teachers and new fellow students. Again, accent on the word NEW. However, the crucial challenges are academic as students encounter a more developmental and demanding level of critical thinking in their young adolescent lives. How can we, teachers, counselors, and administrators, help our students not only to adjust to this new academic and social setting, but more importantly, to thrive as middle school students?

A recent study by Lemberger-Truelove et al., 2021, demonstrated the importance of counselor assistance in the academic development of middle school students. "Generally speaking,

DOI: 10.4324/9781003597292-6

when school counselors engage students in direct services, results indicate desirable outcomes in a variety of areas including students' social and emotional development, academic performance, and college or career pursuits" (Griffith et al., 2019; Whitson et al., 2011). Lemberger-Truelove et al. recognize that these results are encouraging, but that there is a paucity of literature on this topic. At the same time, one of the problems confronting counselor assistance to middle school students is that counselors are encumbered with tasks that limit their direct intervention with students.[1]

Inspired by the ASCA Mindsets & Behaviors (ASCA, 2014), this study (Lemberger-Truelove et al., 2021) investigated a combined social and emotional learning and mindfulness-based intervention as delivered by school counselors to students in classrooms and their teachers using consultation practices. The study used a cluster-randomized design at the classroom level, with an ethnically diverse sample of 109 middle school students divided between treatment and delayed treatment groups. This study indicates that students need certain skills for academic success, college and career readiness as well as social and emotional development. Understanding the effectiveness of this study makes a strong argument for counselor assistance as an essential part of middle school life.

Of critical importance to this study are two questions:

1. Do middle school students who participated in the classroom-wide Advocating Student within Environment (ASE) intervention self-report changes in executive functioning and social curiosity?
2. Do middle school students who participated in the classroom-wide ASE intervention perform differently on a standardized academic achievement in the areas of math, science, English, and social studies?[2]

In regard to Question 1, Lemberger-Truelove et al. found significant growth in executive functioning skills. Teachers saw this improved ability of students in the treatment protocol to perform the essential tasks of executive functioning. In particular, teachers saw significant growth for all students—treatment group and delayed treatment group—in the ability to "inhibit" distracting stimuli, a highly essential growth pattern for adolescents. Similar results were found for stress tolerance and social curiosity. For stress tolerance, all students improved, but students in the treatment group experienced greater growth than students in the delayed treatment group. Students in the treatment program had positive change in their social curiosity while students in the delayed treatment program declined in their social curiosity.[3] Most importantly, students in the treatment group were seen by their teachers as better able to perform the tasks of executive functioning.

Question 2 also found positive responses. Student performance on achievement tests for English, science, and social studies "demonstrated that students who were exposed to the intervention improved in their academic achievement.... for mathematics, all students made gains in their pre to post test scores... but the intervention between time and intervention status was not significant."[4] What is clear in this study is that all students benefited from this intervention program designed to make students aware of executive functioning and how it impacts so much of their young lives.

According to another study of early adolescent development in the middle school years (Wigfield et al., 2005 p114), educational and developmental researchers frequently measure self-concept as a sense of competence or the ability to participate in different activities. Young children have "clearly separable dimensions of self-concept.... Quite early on, children have differentiated views of themselves and so during early adolescence these beliefs are defined clearly."[5] Children have already developed beliefs about their competencies and abilities in different subject areas in the same way that they have acquired beliefs about exhibited abilities in particular sports or other activities. Self-esteem refers to an individual's sense of overall worth or value as a person. However, particularly for early adolescents, self-esteem may be very vulnerable due to multiple changes: puberty, school change, and changes in family and peer relations.

Identity is a broader term than self-concept and self-esteem, referring to the individuals' sense of themselves and their psychological reality, including beliefs and attitudes about themselves.[6]

What is important for educators to understand is that while students may have these pre-established beliefs, young adolescents are still in a formative stage. Now is the time to assist them in this key stage of life, while at the same time, recognizing that for some students, stronger efforts may be necessary to help them feel more confident about themselves: personally, academically, and psychosocially.

Also, as noted in Chapter 2, middle school is a formative stage of adolescent life where self-concept, self-esteem, and identity are critical; in other words, a time of identity formation. Identity is a term broader than either self-concept or self-esteem, referring to individuals' general sense of themselves and their psychological reality that includes many different beliefs and attitudes about the self (Wigfield and Wagner 2005). Indeed, "identity formation involves the successful negotiation of a variety of activities and relationships…including school achievement, social relations with others, and development of career interests and choices, along with a great deal of exploration of different activities and roles."[7] Clearly, life for middle school children consists of many dynamics ranging from academic achievement to social relationships and participation in extracurricular activities and sports among others, making this a truly formative period of identity formation in their young lives.

Furthermore, from the perspective of entering middle school students who are accustomed to more consistent teacher contact in elementary school, there is less interaction with their teachers as they enter middle school. For most middle schoolers, the student-teacher interaction occurs in the subject classroom and perhaps in a once weekly after-school extra help session with given teachers or in an extracurricular activity that a teacher may moderate. Often, middle school students report that the quality of their relations with teachers has declined from what it was in elementary school.[8] This lack of teacher interaction may heighten students' apprehensions as they strive to find their ways in this new phase of their education.

What both studies present is a strong argument for a consistent methodology designed to assist middle school students on their K–12 journeys. This early adolescent period is critical in developing the necessary life skills that are rooted in executive functioning. Indeed, "early adolescents' increasing cognitive sophistication means that school counseling programs can be geared to this higher-level thinking, helping adolescents to focus on the future and the possibilities it may hold."[9] As middle schoolers are moving into more abstract thinking, their executive functioning abilities need guidance and direction that will result in academic achievement as well as the successful attainment of the many other endeavors that students choose to pursue. At the same time, identity formation is developing during this highly critical time of young adolescence that impacts their self-perceptions about their abilities for future endeavors. Clearly, this is a highly opportune time, perhaps a necessary time, for educators to actively assist middle school students in this all-important stage of their young lives.

So, how to best assist middle school students at this truly formative stage of their young lives? What we present here are two possible formats for providing this essential assistance to middle school students. As coauthors, we have experienced both models to be effective experiences for middle school students. Educators must choose which format would work more successfully in their schools. The advisory system or the group counseling program is not designed to replace the ongoing individual counseling session with a school counselor, social worker, or school psychologist.

Model 1: The Advisory System

What we suggest is the creation of an ongoing advisory system to assist students in these critical middle school years. Using an advisory system can help middle schoolers understand themselves as students, leading them to a broadening view of their aspirational goals as they learn how to meet academic challenges and balance the demands of social and extracurricular life.

Who Are the Advisors?

In this model of an advisory system, teachers, counselors, and administrators all become advisors and follow a common path in addressing the needs of middle school students. Advisors meet once a week/cycle for a minimal 40-minute period with their advisory group and follow a shared agenda that is presented in the following chapters of Part II. An alternative model of the advisory system, with teachers, counselors and administrators as advisors, is the bi-weekly/cyclical advisory program that includes both group sessions as well as individual meetings with students. The common agenda of both models of the advisory system includes helping students to understand what effective executive functioning means from the development of constructive goal setting to strong study skills and time management abilities. As students progress, the shared agenda moves into an initial career exploration and college exploration.

We, as coauthors, experienced the effectiveness of this all-inclusive advisory system as part of a federal grant aimed at assisting schools in improving overall school performance. While the aim is to assist students, one of the collateral benefits of this system is to broaden the relationships of all involved as administrators, teachers, and counselors engage in helping students become more aware of executive functioning and how it impacts their lives in multiple areas from academics to sports to extracurricular activities. As the program progresses, advisors will also experience the excitement of middle school students exploring possible careers and colleges. In other words, advisors are helping students to develop the skills and habits that are necessary for academic success as well as sparking the interests in their future lives—both at the heart of college and career readiness.

Essential to the success of the advisory system is the willing participation of all advisors—teachers, administrators, and counselors. A principal, assistant principal, or head of the guidance department may be assigned the responsibility of overseeing the program. A key responsibility is conducting regular meetings of all advisors in which the shared agenda is presented, with follow-up meetings to discuss progress. What follows in Part II of this book is a step-by-step guide to the shared agenda that is briefly presented below.

We, as coauthors, had the rewarding experience of seeing how young minds can understand and incorporate a working form of executive functioning into their lives. At IN-Tech Academy, a 6–12 New York City Public School in the Bronx, NY, we created an effective Advisory System in which teachers, counselors and administrators followed a shared agenda in their weekly meetings. As the second quarter ended, we attended a sixth-grade advisory session. Here the advisor led a discussion of whether grades had improved over the second quarter. As the advisor went subject by subject, overwhelmingly, these sixth graders had improved their grades. When the advisor asked what had contributed to this success, the answers made perfect sense. One student responded: "I wrote down my homework"; a second said, "I remembered to bring it to school," while a third student proudly stated that he had actually handed in the homework to the teacher! While this may seem to be a normal course of action, life in middle school may be far from the expected norm. These students may not be thinking in terms of executive functioning, but their actions revealed that this term was operative, and hopefully, a growing concept in their young minds.

Model II The Group Counseling Program

For schools where the implementation of an advisory system may not be possible, counselors take on the role of the advisor. Here the head or assistant head of the guidance department assumes the leadership role. Similar to the advisory system, in the group counseling program, counselors meet with their group once a week/cycle. Again, as in the advisory system, the group counseling program may also be scheduled in the bi-weekly/cyclical format that includes both group sessions and individual meetings. Just as in the advisory system, counselors should meet regularly to discuss middle school students' progress as they further develop executive functioning and study skills to career and college initial exploration. Both models use the same shared agenda that is presented in the following chapters of Part II. Again, we as coauthors have successfully used the group counseling structure in our careers.

Both the advisory system and the group counseling structure achieve positive effects. Educators should decide which format might be more effective in their individual school environment.

The Shared Agenda of the Advisory System and the Group Counseling Program

Chapters 5–10 of Part II present a systematic approach to helping middle schoolers become proactive students who will become increasingly aware of executive functioning: what it means and how it impacts not only their study skills, but also their overall lives. Students will also learn the importance of setting goals and the how behind achieving them. Intrinsic to this approach is a developing awareness of career possibilities and an initial venture into college exploration. Part II concludes with examining secondary school options and ideas for enriching summer activities that may broaden students' exploration of areas of interest to them.

The following brief chapter summaries highlight essential work to be done in advisory systems and/or group counseling sessions, i.e., the shared agenda of both models. In the actual chapters, the material is presented in the voice of the advisor/counselor speaking to students.

Chapters 5 and 6 are specifically geared to sixth grade middle school students.

Chapter 5: Helping Middle School Students Understand Executive Functioning

In this chapter, we discuss the importance of student articulation of goals as they enter middle school. We also explore how students' understanding of Executive Functioning may help them to enhance their ExF Factor in goal-driven tasks. In addition, we provide advisors/counselors with charts and exercises designed to assist students in identifying their academic and extracurricular goals and the means of achieving these aims. In this way, students are actively engaged in understanding and achieving their aspirations. Importantly, we also address the negative effects of constant monitoring of social media sites and doomscrolling, both of which have dramatically risen over the last few years among middle school students. Finally, follow-up exercises to evaluate to what extent students have effectively achieved their goals will assist advisors/counselors and students in assessing the best strategies to reach goals and to understand what changes may be needed to achieve greater success.

Chapter 6: Developing Enhanced Study Skills for Academic Success

This chapter provides a step-by-step guide to developing enhanced study skills. Using learning style and study skills inventories, advisors/counselors will deepen students' awareness of the learning process and their own growth in the transition process. In this chapter, we also discuss how technology has changed the ways in which students learn and the various ways that students must present their completed work. Included here are the traditional homework assignment books and presentation methods as well as the increasing use of learning management systems. Also addressed are the negative impacts of using ChatGPT in writing essays. As middle schoolers encounter not only higher academic demands but also new methodologies, this chapter provides a structured approach to help them succeed in this all-important stage of their lives.

Chapter 7: Using Technology in Middle School

In this chapter we examine how technology is transforming the modern classroom. Essential to this discussion is how students, with the help of advisors/counselors, can use innovative technologies to enhance their study skills and goal achievement.

1. Laptops and tablets are essential for facilitating interactive and portable learning environments.
2. Interactive SMARTboards/Whiteboards increase student engagement by incorporating multimedia resources and interactive activities.
3. Learning Management Systems (LMS) such as Google Classroom and Schoology are crucial for overseeing student information and enabling online education.

Included is a presentation of Google Classroom and Schoology, and how these learning management systems impact homework completion and student planning.

Chapters 8 and 9 relate more to 7th and 8th grade students.

Chapter 8: Career Exploration in Middle School as a Motivational Technique

This chapter explores the time-honored question: "What do you want to be when you grow up?" In exploring the answers to this question, students will use various websites designed to help students explore careers through interests, values, and talents assessment quizzes. Particular attention is given to the ACT website which allows students as young as 11 years old to take career inventories that lead to the Career Map. (To access the resources of the College Board, students must be age 13, consequently, limiting its use by middle school students.) Advisors/counselors will assist students in an exploration of possible careers and help them to understand the day-to-day work that a particular career involves. Through these activities, students will be exposed to a wide range of careers that they may not yet have encountered. Using charts and exercises, students will then delineate the educational background necessary for a particular career. Advisors/counselors may then encourage students to use this information to motivate them to achieve greater academic success. Also included is a discussion of how Career Fairs may be beneficial to middle school students, as well as a questionnaire that students may use to interview a person in a career field of interest to them.

Chapter 9: College Exploration in Middle School

Using various websites, students will begin an initial search of colleges. Students who used the career exploration of the ACT will now investigate colleges available on the ACT's Explore Schools section of the website. Students who used another website in career exploration should now use the ACT website to explore colleges. (The College Board's resources are available to students age 13.) Advisors/counselors will be able to assist in this exploration by using the charts and exercises that are provided in this chapter. Essential to this exploration is the vocabulary of post-secondary education from majors and minors to liberal arts colleges and universities. Also included in this chapter is an introductory overview of financial aid options as well as specific programs that many students may find helpful to achieve their college aspirations.

Chapter 10: Articulation to High School: Finding Strong Matches in an Expanding World of High School Possibilities
Every fall in the United States, eighth graders, and in some cases sixth or seventh graders, begin to explore and apply for high school placements for specialized schools or regular public, private, or parochial schools. With the explosion of charter schools, specialized schools, and the closing of poorly performing schools nationwide, getting into high school, knowing the choices, and most importantly, being prepared to apply for a particular high school, has become a crucial part of the work of middle school counselors, teachers and administrators. This chapter will provide guidance in understanding the process, including the different types of schools and the academic advisement needed to enable middle school students to have choices of high schools where they can matriculate and be successful. The chapter will also include ideas for specialized programs and summer activities that facilitate the development of skills and confidence.

Ongoing Discussion of Student Progress with Subject Teachers in the Advisory/Group Counseling Systems

The work of both models, the advisory and the group counseling, is designed to help middle school students develop a foundational basis of strong study skills and habits that the College and Career Standards describe as the requisite benchmarks for future success. The shared agenda of the advisory/group counseling systems is designed to be presented over multiple sessions throughout the middle school years. Discussing how well students are performing in subject areas is essential to this process. In the advisory system, teachers, counselors, and administrators are working with students, and it is necessary that the subject teachers (some of whom may be the advisors) understand what their students are learning in these advisory sessions. Likewise, in the group counseling model, periodic meetings with subject teachers are especially important as teachers come to understand what their students are learning about executive functioning and how its four components may be applied to their particular subject areas. These periodic meetings, both advisory and group counseling, are a perfect opportunity for individual teachers to share their observations of how students are performing in their classes. Such an exchange may help all—teachers, counselors and administrators—in advancing the shared agenda, designed to assist students in this all-important stage of their K–12 journeys.

Furthermore, meetings with subject teachers in both models help to achieve the aim of building foundational study skills and habits. Particularly, in presenting executive functioning and how it impacts academic success, the advisors/counselors meetings with the subject teachers can only enhance student progress. These meetings are especially important in the early stages of the advisory/group counseling sessions where middle school students are learning about executive functioning. As the advisors and counselors are helping sixth graders understand executive functioning and its correlating components, subject teachers' responses as to the extent that students are implementing this work into their classroom experience are highly valuable. At the same time, advisors and counselors are not trying to replace teacher expectations in particular classes. For example, as advisors and counselors move into the executive functioning component of organization, they should be aware of the subject teacher's expectations regarding the methods of organization in particular subjects. These meetings are an opportune time to share any specific organizational methods or note-taking styles that individual teachers prefer their students to use. For example, the methodology for note-taking and organization in a science class may differ from an English or social studies class. What we are trying to ensure is that all middle school students understand how strong executive functioning is essential to academic success.

As the shared agenda of the advisory/counseling system progresses, input from the subject teachers may be very helpful. For example, students exploring careers spark an excitement about their future lives. Many schools have Career Fairs where people engaged in various career endeavors speak to eager middle school students. Professionally, we have both seen how students become positively engaged in this kind of activity. Additionally, members of the faculty in various academic areas shared the excitement as they helped to find friends and family members in a broad range of careers. Even if a Career Fair is not possible, students using the ACT website (the College Board requires students to be 13 years old to access the website) are able to learn about the many career possibilities in the fast-changing 21st century global economy. As the sixth–seventh grade advisory/group counseling agenda proceeds to an initial college exploration, subject teachers may share their own experiences. Particularly as eighth grade students are considering specialized high schools, subject teachers may be helpful in the advising process. As middle school students transition to high school, our aim is to ensure that they have a strong foundational basis that will aid them in meeting the challenges of secondary school as well as having an openness to the many career and college possibilities in their futures.

*Please note that the agendas for second semester advisory/group counseling sessions for eighth graders are presented in **Chapter 15** in the **Helping Students in the Transition to High School (2nd semester 8th grade students)** section.

As the Wigfield et al., 2005 and the Lemberger-Truelove et al., 2021 studies demonstrate, middle school students clearly benefit from the consistent intervention of counselors. These young students are at a pivotal point in their lives. Now is the appropriate time to assist them as they develop their abstract thinking through an understanding of executive functioning and its related components. As students establish goals, they will experience the effectiveness of goal accomplishment through their developing executive functioning abilities. Academically and personally, this process can only enhance students' belief in their own abilities to take on tasks successfully as they develop in middle school. As they progress in developing strong and effective study skills and habits, again, these middle schoolers will experience greater academic success, leading to an opportune time for them to explore careers and delve into an initial college search.

As educators, we have the opportunity to assist these young adolescents in a constructive manner that will broaden their vision of themselves now and in the future. Using either the advisory system or the group counseling program provides the framework for a consistent, ongoing approach in helping middle school students grow academically and personally in these formative years. We, as coauthors, have seen the effectiveness of both approaches. What is essential is to incorporate this developmental help into the school program so that middle school students may experience the positive effects of a consistent approach to enhance their understanding of themselves as students and to broaden their sense of future possibilities in their young lives.

Notes

1 Lemberger-Truelove, Matthew et al. "Growth in Middle School Student's Curiosity, Executive Functioning and Academic Achievement." *Professional School Counseling*, May 2021, Vol. 24, No. 1b, Special Issue: Educator–Counselor: A Nondual Identity for School Counselors, p. 1.
2 Lemberger-Truelove, Matthew et al. "Growth in Middle School Student's Curiosity, Executive Functioning and Academic Achievement." *Professional School Counseling*, May 2021, Vol. 24, No. 1b, Special Issue: Educator–Counselor: A Nondual Identity for School Counselors, p. 2.
3 Lemberger-Truelove, Matthew et al. "Growth in Middle School Student's Curiosity, Executive Functioning and Academic Achievement." *Professional School Counseling*, May 2021, Vol. 24, No. 1b, Special Issue: Educator–Counselor: A Nondual Identity for School Counselors, p. 4.
4 Lemberger-Truelove, Matthew et al. "Growth in Middle School Student's Curiosity, Executive Functioning and Academic Achievement." *Professional School Counseling*, May 2021, Vol. 24, No. 1b, Special Issue: Educator–Counselor: A Nondual Identity for School Counselors, p. 5.
5 Wigfield, Alan, et al. "Early Adolescents' Development Across the Middle School Years: Implications for School Counselors." *Professional School Counseling*, December 2005, Vol. 9, No. 2, Special Issue: Focus On Middle School Counseling, p. 114.
6 Wigfield, Alan, et al. "Early Adolescents' Development Across the Middle School Years: Implications for School Counselors." *Professional School Counseling*, December 2005, Vol. 9, No. 2, Special Issue: Focus On Middle School Counseling, pp. 113–114.
7 Wigfield, Alan, et al. "Early Adolescents' Development Across the Middle School Years: Implications for School Counselors." *Professional School Counseling*, December 2005, Vol. 9, No. 2, Special Issue: Focus On Middle School Counseling, pp. 113–114.
8 Wigfield, Alan, et al. "Early Adolescents' Development Across the Middle School Years: Implications for School Counselors." *Professional School Counseling*, December 2005, Vol. 9, No. 2, Special Issue: Focus On Middle School Counseling, p. 115.
9 Wigfield, Alan, et al. "Early Adolescents' Development Across the Middle School Years: Implications for School Counselors." *Professional School Counseling*, December 2005, Vol. 9, No. 2, Special Issue: Focus On Middle School Counseling, p. 113.

5

Helping Middle School Students Understand Executive Functioning

Leaving the more familiar world of elementary school, many middle school students may truly feel that they are entering a new realm of education that will make greater demands that may initially feel overwhelming. Middle school will challenge students academically, socially, and emotionally. An effective approach to helping students meet these challenges is to have them establish goals and to assist them in finding an effective means of achieving their articulated objectives in academic performance, in participation in extracurricular activities, and in the sports that they enjoy. While this approach may seem quite logical, what students will learn in this process is a new term to them: executive functioning—what it means and how they may improve their ExF factor.

What is important for us to know as educators is that adolescents' brains are experiencing a highly developmental period. The National Institute of Health's ongoing *Adolescent Brain Cognitive Development Study* has been following 12,000 youths over a ten-year period at 21 research sites across the country. This study is exploring the environmental, social, genetic, and

DOI: 10.4324/9781003597292-7

biological factors that affect brain and cognitive development, behavior, and health.[1] Recent findings in this research study have found that the brain undergoes significant remodeling during the teen years, reaching maximum density in the cortex before maturing fully in the mid-to-late 20s. Furthermore, the limbic system (reward center) develops faster than the prefrontal cortex (responsible for planning and decision-making), creating an "imbalance" that often leads to increased risk-taking, emotional intensity, and peer-focused behavior. The study has also found the teen brain is highly adaptable, allowing for strengthening of circuits through learning, exercise, and "positive" risks like joining teams, but also making it vulnerable to negative environmental impacts.[2] Clearly, as the ABCD Study shows, adolescence is a period of dramatic brain development. It is essential for educators to understand that middle school students are in this period of mental and emotional development that is open to new ideas and approaches to achieving academic and personal success. Particularly, as the ABCD Study notes that the teen brain is highly adaptable, allowing for strengthening of circuits through learning, the middle school years are a crucial period to establish the study skills and habits that are the foundational base of the College and Career Readiness Standards.

Furthermore, as noted in the earlier Wigfield et al. study, early adolescent thinking changes in significant ways. Adolescents increasingly engage in abstract thinking that allows them to consider both the hypothetical as well as the real. They also engage in more sophisticated and elaborate information processing strategies that reflect on themselves, and in particular, their complicated problems. In addition, scholars now believe that "adolescents' increasing ability to organize and reflect on the information that they have allows them to engage more readily in these higher order thinking processes."[3] Adolescent students are clearly capable of abstract thinking, leading to a higher level of critical thinking that will be demanded academically in middle school and beyond. As middle school students experience heightened academic demands, understanding executive functioning and its related components is essential for academic and personal success. Most importantly, as adolescents strengthen their executive functioning abilities, they should see a clear correlation in their achievement of academic and personal goals.

How do we as educators explain executive functioning and its necessary components to middle school students? In a recent study, Lemberger-Trulove et al. referred to the components of executive functioning as a student's ability to inhibit, shift, plan, organize, and monitor tasks.[4] In other words, the capability of a student to avoid distractions, to move from one subject to another, to plan and organize effectively, and to monitor and/or complete given assignments. In the following sections of this chapter, these terms will be introduced to describe effective executive functioning with particular emphasis on a student's ability to inhibit distractions. Indeed, while we are speaking about middle school development, the components of executive functioning are essential for all stages of life. Middle school students are quite literally on the brink of a new, more demanding stage in their K–12 journeys.

Of special importance of the four executive function components is the ability to inhibit distractions. While the world has always found ways to distract people, the widespread use of the internet and cell phones has raised serious concerns for middle school students. The adolescent world is increasingly filled with distractions. As noted in Chapter 1, one distraction that is of growing concern for middle school students is the frequent, if not constant, use and visiting of social media sites. According to The Annie E. Casey Foundation, current levels of teen social media use are concerningly high. "Virtually all teens (96%) ages 13 to 17 use the internet daily, with nearly half saying they are online… Of the five most commonly used social media apps—Youtube, TikTok, Instagram, Snapchat, and Facebook—a third of teens report using one or more of these 'almost constantly.' While most US social media platforms require users to be at least 13 years old, an estimated 64% of kids ages 8 to 12… use YouTube and TikTok every day."[5] When speaking of distractions, social media is a prime concern in the lives of preteens and teens. An important aspect of executive functioning is the ability "to inhibit" such distractions, making this a topic of concern in this chapter.

Another worrying activity is the use of the internet by middle school students who engage in doomscrolling, a word added to the Merriam-Webster Dictionary in 2020 and defined as spending excessive time online scrolling through news or other content that makes a person feel sad, anxious, angry, etc.[6] While some time spent looking online may be normal for many people, excessive doomscrolling and mindless consumption of digital content can have serious consequences. Particularly for middle school students, doomscrolling may have significant negative effects. According to Nationwidechildrens.org, "In today's digital age, kids and teens are spending more and more time scrolling through their phones, often without much thought or purpose. This behavior, while seemingly harmless, can actually be detrimental to both their minds and bodies."[7]

Possible results of doomscrolling include:

- **Reduced attention span**: Scrolling through bite-sized content constantly trains the brain to seek quick, immediate rewards. As a result, it becomes harder to focus on more complex tasks, whether it's homework, reading a book, or engaging in thoughtful conversation.
- **Diminished critical thinking**: Doomscrolling tends to erode deep thinking. When kids are bombarded with surface-level information and breaking news headlines, their ability to critically analyze and reflect diminishes. Instead of pondering a topic in depth, they become accustomed to just skimming the surface.
- **Exposure to misinformation or harmful information**: Not everything we see and read online is true or right—even adults can have a hard time making sure they're getting accurate and harmless information sometimes. For example, when an influencer vapes, it doesn't mean that vaping is "good" or healthy, but often youth may perceive it as cool or good when they see others using it.
- **Emotional fatigue**: Constant exposure to distressing news can take a heavy emotional toll. The flood of information can leave kids feeling overwhelmed, anxious, and unable to process their own feelings.

- **Social isolation**: Social media often lacks the depth and authenticity of real-world connections, possibly leaving kids feeling more isolated despite being "hyperconnected."
- **Physical symptoms**: Long hours of doomscrolling can lead to physical effects like headaches, nausea, difficulty sleeping, elevated blood pressure, and even a low appetite.

While doomscrolling may be a relatively young phenomena, its impact has dangerous consequences for young minds, possibly a decline in cognitive abilities such as attention, memory, and mood.[8] Clearly, the impact of doomscrolling has potentially serious, even precarious consequences for middle school students. As educators, we are trying to help adolescents develop their cognitive abilities and enhance their executive functioning; doomscrolling is not only an interrupter, but it is also a dangerous distraction with possible detrimental consequences.

Beginning in the 2025–2026 school year, the New York City Public School System, the largest school district in the nation, updated its policy regarding the use of personal internet-enabled electronic devices, including cell phones in school. This new policy supports the New York State law to create "distraction-free schools". The Chancellor's Regulation A-413 clearly defines this policy:

A personal internet-enabled device is any electronic device not issued by a school or NYCPS program that can connect to the internet, allowing the user to access content online. Examples of these personal devices include cell phones, smartphones, and smartwatches, as well as non-school issued laptops, tablets, and iPads. Also included are MP3 players and game consoles.

When students enter the school building, they must store any devices in school-assigned personal or shared lockers, individual cell phone pouches, or another storage space for the entirety of the day. Individual schools determine storage space and collection and return routines. Devices are collected at the beginning of the day and returned at the end of the day. Teachers are not responsible for collecting devices in the classroom. An added note: backpacks are not considered safe storage. Exceptions to the device policy do exist, primarily for medical monitoring, translation services, and student caregivers, among other exceptions.[9]

All New York City Public School students are issued a school iPad/laptop that is used specifically for schoolwork in the classroom and at home. Content and accessibility on these devices is part of Google Classroom. These devices are not intended for access to social media or personal email; rather they are an extension of the classroom. Chapter 7 on technology presents both Google Classroom and Schoology, the two most commonly used learning management systems in the United States. The New York City Public School System is making a systematic approach to creating a schoolwide environment free of social media visits, doomscrolling, and personal texting or calling during the school day.

As we begin presenting the advisor/group counseling sessions, constant reminders about the importance of focusing on academics and avoidance of distractions are threaded into the presentations. Part II of this book examines the **how** behind increasing academic achievement. In Chapter 5, we aim to help middle schoolers become aware of executive functioning and how it impacts goal achievement. In Chapter 6, we present the four keys to improving students' **ExF factor**: effective planning, organization, time management, and avoidance of distractions—all essential in building the foundational skills for higher achievement. A constant factor in both chapters is the ability to inhibit; in other words, the ability to avoid distractions that draw students away from their work with particular emphasis on cell phones and online access to social media sites. As educators, once again, we have the opportunity to stress the importance of developing executive functioning not only in academic performance, but also in personal activities such as extracurricular clubs, sports, and family responsibilities.

Memo to Advisor/Counselor

As students enter middle school, they must learn more advanced study skills and become more aware of how the completion of their assignments is related to an ongoing development of Executive Functioning—their ExF factor. Students entering this next level of their education will experience multiple challenges. Teachers, counselors, and advisors can provide invaluable assistance to middle schoolers as they encounter increased academic demands. While individual subject teachers provide guidelines for classroom participation and homework, the advisory/group counseling systems provide an essential framework for student growth. As noted in *The Teachers College Survey Research Initiative,* College and Career Readiness Standards are benchmarks that define the knowledge, skills, and abilities students need to succeed in both college and career pathways after high school. The journey to achieve these standards continues throughout K–12. [10] Our focus is to assist middle school students in this stage of their journey to meeting the benchmarks essential for future success. Helping middle school students understand executive functioning is a necessary component for their success. As stated above, students must comprehend the components of executive functioning: the ability to inhibit, to shift, to plan, to organize, and monitor tasks. In other words, the ability of a student to avoid distractions, to move from one subject to another, to plan and organize effectively, and to monitor and/or complete given assignments. Students will vary in their executive functioning, but all middle schoolers will benefit from seeing how improved executive functioning increases academic achievement. As they truly engage in improving executive functioning components, students are moving forward in establishing the foundational study skills and habits that meet the College and Career Readiness Standards, an essential goal for middle school students in their K–12 journeys.

A side note: As an educator, this coauthor Kathleen Costantini has presented executive functioning and its related components to middle school students who quickly came not only to understand this essential term, but also to incorporate the components (with guidance, of course) into their young academic lives. What I learned here was that parents also needed to understand the terminology being used in the classroom. (Please see Chapter 14 for Parents Meetings on this topic.)

The following information is presented in the voice of the advisor/counselor speaking to middle school students after a few days or a week of school. This section may be adapted to the particular needs of students within a district.

Welcome to Middle School!

As we start classes, you may be experiencing some challenges: meeting new people, both students and teachers, changing classrooms for subjects, using lockers, eating lunch at different times on various days, new afterschool clubs and sports—even how you travel to school may be different! Now that you have spent a few days in middle school, let's complete Exercise 5.1.

EXERCISE 5.1 First Impressions of Middle School

List 3–5 changes you have experienced in middle school.	*Evaluate each change: Do you like or dislike the change?*	*How can you best adapt to this change?*
1.		
2.		
3.		
4.		
5.		

Once students have completed Exercise 5.1, divide students into small groups to share their responses. After sufficient small group discussions, the advisor/counselor returns to a large group discussion where students may share their comments. Often just knowing that other students have similar concerns boosts students' confidence.

In a following session, advisors/counselors may move onto discussing goals with students. Again, this presentation is in the voice of the advisor/counselor.

At today's meeting we are going to discuss goals and how to achieve them. Now that you are in middle school, academic demands increase. At the same time, you will have the opportunity to join new activities. Your middle school offers a range of clubs and sports teams. Don't hesitate to try something new!

Let's take the time now to think about what you want to achieve this year. As a middle school student, you are at the threshold of a more demanding academic experience with the possibility of participation in new activities and sports. Let's think about your goals for sixth grade!

The advisor/counselor may want to provide a list of extracurricular activities prior to presenting this exercise. Many schools have a Club Fair to introduce new students to activities; if this is the case in your school, time Exercise 5.2 according to the school calendar.

Now let's complete Exercise 5.2. Here you need to think about your goals for the first quarter/semester, and most importantly, the HOW behind reaching your goals! It's always best to think about your aims and to plan a way of achieving your goals.

Once students have completed Exercise 5.2, advisors/counselors should conduct a group discussion of student goals and planned methods of achieving them.

Advisors/counselors should also have copies (digital or paper) of students'—completed Exercises 1 and 2. These completed exercises are essential to discussing first quarter/semester grades and improved methods to achieve goals. While students have established goals in several areas, the primary concern of this chapter and its related sessions with sixth graders, is helping students to learn what is necessary for successful achievement of academic goals.

EXERCISE 5.2 Establishing Goals and How to Achieve Them

Goals	*How To Achieve*
Academic Achievement	
Extracurricular Activities	
Sports	
Other	

Note to the advisor/counselor: The Sallie Mae Fund's Kids2College Program tells middle school students: "It might sound like everything you need to do to get ready for college happens in high school. This isn't the case! There are things you can do right now as a middle school student to put yourself on track for college." How students learn to approach both homework and studying are key in developing a consistent methodology for not only completing assignments, but also for pursuing areas of interest to them. What students learn in middle school becomes the foundational skills and habits that will help them to achieve success as they proceed on their K–12 Journey.

Executive Functioning and the ExF Factor

Note to the advisor/counselor: According to the Cleveland Clinic, brain development in adolescents allows them to "deal with abstractions, test hypotheses and see infinite possibilities...the frontal cortex continues maturing until people reach their mid to late 20s... This area of the brain controls executive functions such as planning, prioritizing and controlling impulses."[11] Developmentally, middle school students are capable of understanding executive functioning and its related components—effective planning, organization, time management, and avoidance of distractions. In the following sessions, advisors/counselors should present the term Executive Functioning and help students to understand how they may improve their ExF factor! Understanding Executive Functioning and their own ExF factor is essential for middle school students as they work to achieve their goals. Students will vary in their executive functioning abilities. Here, we are using the **term ExF Factor** to personalize where students are in this process.

Again, in the voice of the advisor/counselor....

Now that you have established goals, let's look at two essential terms: **Executive Functioning and your ExF Factor.**

What is **executive functioning?** Think of it as a developing part of your brain that helps you to plan, organize, manage time, and avoid distractions. As a middle school student, you are at an important stage in this development. How can executive functioning help you to achieve your goals? While we are talking about academic goals in these sessions, the same skills and abilities can aid you in accomplishing goals in extracurricular activities and sports. You need to prioritize and plan your work, to organize your thoughts and to complete your assignments effectively, and make efficient use of your time. Also, good executive functioning will not allow distractions to prevent you from completing your work. Doomscrolling and constant viewing of social media sites are distractions! When you are studying, turn off your cell phone! Time to focus only on your work! You will have time to talk or text with friends during breaks. You may want to coordinate break times with friends so that you don't interrupt each other's study time.

Another Way to Think of Executive Functioning: Four Essentials Needed to Achieve Goals

- Effective Planning
- Organization
- Time Management
- Avoiding Distractions

Executive functioning is a driving force in achieving goals. Your **ExF factor** focuses on how well you are able to complete tasks and assignments. Understanding your **ExF factor** can help you improve how you plan, organize your work and manage your time more effectively. Avoiding distractions is also essential in developing your **ExF factor!** Looking at your cell phone or texting during your study time is a distraction. Plan a time when you may talk or text with friends—not during your designated study time!

Now, complete Exercise 5.3. Remember, look at yourself honestly! This exercise is not about right or wrong answers—it is about helping you to understand how you may better achieve your goals.

After completing this exercise, note how many times you marked a No or Sometimes. These answers are areas where you can improve your executive functioning to become a better student. Remember, even if you marked No or Sometimes multiple times, it is not a wrong answer! It is a signal telling you where you may improve your **ExF factor!**

EXERCISE 5.3 Executive Functioning

Question	*Yes*	*No*	*Sometimes*
Do you look at your daily assignments and plan how you will complete them?			
Do you prioritize your assignments by thinking about what will take the most time or effort?			
Do you think about what is the most important task for you to complete?			
Do you use a planner to write down assignments?			
Do you break long-term assignments into smaller parts?			
Do you correctly estimate how long it will take you to complete an assignment?			
Do you have a nightly study schedule that you try to follow?			
Do you organize your schoolwork neatly in a binder or notebook?			
Do you review on a regular basis, not just before a test?			
Do you keep a record of your grades in each subject?			
Do you include the demands of your job, sports and other activities in your planning to complete your homework?			

Again, advisors/counselors should keep a copy of students' Exercise 5.3 (digital or print) for future comparison. What is important in this exercise is that middle school students are engaging in self-analysis, which is a necessary step in understanding their own executive functioning and how it may be improved. For many middle school students, this analysis may be occurring for the first time.

A side note: As an educator, this coauthor Kathleen Costantini encountered a student who more than likely had never heard of the term executive functioning but exhibited the very definition of a high ExF factor in what for me was a memorable way. I had previously taught at Regis High School, an all-scholarship school in New York City, and continued to interview students whose admission scores and grade school transcripts were in the range of acceptance for the school. When I asked eighth grader, Pablo, what kind of extracurricular activities or sports that he enjoyed, he responded quietly that he did not have any outside activities. He appeared to be a bit uncomfortable, perhaps feeling that he was lacking qualifications. Trying to put him at his ease, I asked him to tell me about what he did after school. Pablo then proceeded to explain that both of his parents worked and that he was responsible for taking his younger brother and sister to the local Catholic school that they all attended. At the end of the school day, Pablo brought his siblings home, gave them a snack, and then supervised their homework. At 6:00 p.m., he then put the meal that his mother had prepared on the table for dinner. After eating and then cleaning up, Pablo then began his nightly homework, still working when his parents arrived home around 10 p.m. Of course, Pablo had plenty of activities! His ExF factor was extremely high, and yes, he was admitted to Regis High School.

Again, in the voice of the advisor/counselor.....

So, how can you improve your ExF factor? Remember that good executive functioning is based on:

- Effective Planning.
- Organization.
- Time Management.
- Avoiding Distractions.

Think of these terms as the four keys to higher achievement! In our future meetings, we will explore ways of improving each of these keys as we explore how they are linked to improved executive functioning. As we look at each key to improving your ExF factor, we will examine a step-by-step method to help you to plan, organize, mange time, and avoid distractions.

At the end of the first quarter/semester, counselors/advisors should have students complete Exercise 5.4 Evaluating Goals and Achievements of First Quarter/Semester.

After completion of Exercise 5.4, divide students into small groups or conduct one large group discussion to evaluate their assessment of goal achievement, particularly in the area of academic achievement with emphasis on students' analysis of

EXERCISE 5.4 Evaluating Goals and Achievements of First Quarter/Semester

Goals	*How did I do?*	*What were my methods to achieve my goals?*	*What can I change to improve?*
Academic Achievement			
Extracurricular Activities			
Sports			
Other			

how they may improve methods to achieve their goals. Again, advisors/counselors should keep a copy of this exercise (digital or print) for future reference.

Again, in the voice of the advisor/counselor

Now that you have an understanding of executive functioning, your own **ExF factor** and how they relate together to achieve goals, we will move on in future sessions to the **HOW** involved in effective planning, organization, time management, and avoidance of distractions.

A Final Note

As students establish goals, especially in academic achievement, what is important for them to know is that strong executive functioning abilities are essential for goal accomplishment. While middle school students may easily understand that becoming a skilled baseball player takes much practice, achieving academic goals also takes continued practice in planning, organizing, managing time, and avoiding distractions. As we move into an exploration of the four keys to enhancing executive functioning, middle school students will expand their perceptions of goal achievement not only in academic performance, but also in extracurricular activities and sports among other endeavors. Now is the perfect time to assist adolescents as they strive to develop strong executive functioning abilities.

Notes

1 "Adolescent Brain Cognitive Development Study (ABCD Study)." *National Institute of Mental Health (NIMH).* https://www.nimh.nih.gov/research/research-funded-by-nimh/research-initiatives/adolescent-brain-cognitive-developmentsm-study-abcd-studyr
2 "Adolescent Brain Cognitive Development Study (ABCD Study)." *National Institute on Drug Abuse (NIDA), November 2025.* https://nida.nih.gov/research-topics/adolescent-brain/longitudinal-study-adolescent-brain-cognitive-development-abcd-study
3 Wigfield, Alan, et al. "Early Adolescents' Development Across the Middle School Years: Implications for School Counselors." *Professional School Counseling,* December 2005, Vol. 9, No. 2, Special Issue: Focus On Middle School Counseling, p. 113.
4 Lemberger-Truelove, Matthew et al. "Growth in Middle School Student's Curiosity, Executive Functioning and Academic Achievement." *Professional School Counseling,* May 2021, Vol. 24, No. 1b, Special Issue: Educator–Counselor: A Nondual Identity for School Counselors, p. 4.
5 The Annie E. Casey Foundation. "Social Media and Teen Mental Health." *The Annie E. Casey Foundation,* August 10, 2023. https://www.aecf.org/blog/social-medias-concerning-effect-on-teen-mental-health
6 Merriam-Webster Dictionary. https://www.merriam-webster.com/dictionary/doomscroll
7 "Doomscrolling and Brain Rot: What Are They?" https://www.nationwidechildrens.org/family-resources-education/700childrens/2025/01/doomscrolling-and-brain-rot
8 "Doomscrolling and Brain Rot: What Are They?" https://www.nationwidechildrens.org/family-resources-education/700childrens/2025/01/doomscrolling-and-brain-rot

9 "Cell Phone and Electronic Device Policy." *NYC Public Schools.* https://www.schools.nyc.gov/about-us/policies/cell-phone-and-electronic-device-policy
10 "College- and Career-Readiness Standards." *Teachers College Columbia University.* https://sri.tc.columbia.edu/our-framework/college-and-career-readiness-standards/
11 "Adolescent Development." *Cleveland Clinic.* https://my.clevelandclinic.org/health/articles/7060-adolescent-development

6

Developing Enhanced Study Skills for Academic Success

In Chapter 5, we presented middle school students with the term executive functioning and helped them to discern an initial view of their own ExF factor. As middle school students established goals in academic achievement, extracurricular activities, and sports, we again look at the positive effects of counselors working with middle school students on executive functioning. In the Lemberger-Truelove et al. study,

> Analyses found significant intervention effects for the treatment group in students' changes in stress tolerance, social curiosity, executive functioning (i.e., shift, plan and organize, and task monitoring), and academic achievement (i.e., mathematics, science, English, and social studies). Implications of these findings evince how theory-informed school counseling can contribute to important outcomes in educational settings.[1]

Clearly, as this study indicates, counselor intervention positively impacts middle school students in this critical growth period in their young lives.

Furthermore, the December 2024 article, "The Best Way to Study in Jr. High: Creating Valuable Study Skills in Middle

DOI: 10.4324/9781003597292-8

School That Last" stresses the need for study skills development in middle school. This critical time sets the stage academically.

> Workloads start to grow, subjects become more detailed, and children are expected to take on more responsibility for their learning. When students know how to study effectively, they're better equipped to keep up with assignments, handle tests, and manage the demands of multiple subjects without feeling overwhelmed....... Strategies like staying organized, making study guides, or eliminating distractions can help students perform better academically while feeling more in control of their workload.[2]

Truly, the middle school years are crucial in providing students with the necessary study skills and habits that are essential to academic achievement.

Chapters 5 and 6 are designed to assist middle school students in developing the requisite study skills and habits while at the same time nurturing their interest in learning. As noted in Chapter 1, now is the pivotal time for adolescents to develop the foundational basis that is at the heart of the College and Career Readiness Standards.

> These standards (CCRS) are high-quality content standards aimed at ensuring that every student has the skills to compete in today's global, knowledge-based economy, thereby setting them up to pursue the future of their choosing. Standards represent goals for what students should know and be able to do while learning academic content. They differ from curriculum, which provides educators with an outline of what should be taught in the classroom.[3]

What is important is that these standards are not curriculum guides, but rather the emphasis is on what students should know to do and be able to do while learning the academic content of classes. For middle school students, these standards are about learning and developing the fundamental study skills and habits that are the foundation of the adolescent students' future success.

The middle school years are truly a pivotal stage in the adolescents' K–12 journeys. Now that sixth graders have an understanding of executive functioning and their own ExF factor, they are ready to engage in developing enhanced study skills and habits. Indeed, the middle school students' understanding of executive functioning and the four keys necessary to improve their ExF factor will enable them to improve and strengthen their study skills and habits. Using the four keys—effective planning, organization, time management and avoidance of distractions—will enable students to truly develop as students. During this critical stage of adolescent development, we as advisors/counselors can ensure that our middle school students develop the requisite study skills and habits for further learning and academic achievement.

An Overview of the HOW of This Chapter

As advisors/counselors present the four keys to enhanced executive functioning, students' awareness of the learning process and their own growth as students will deepen. Initially, students will take inventories helping them to understand themselves as students as well as their preferred learning style. In this chapter, we also discuss how technology has changed the ways in which students learn and the various ways that students must present their completed work. Included here are the traditional homework assignment books and presentation methods as well as the increasing use of learning management systems. As middle schoolers encounter not only higher academic demands, but also new methodologies, this chapter provides a structured approach to help them succeed in this all-important stage of their lives.

A note to the advisor/counselor: The work of this chapter is substantial as it explores the essence of becoming an engaged, proactive student. Much of the material that is presented here provides the basis of the fundamental study skills and habits that we as educators want to develop in our middle school students. This chapter may take significant time in the advisory or group counseling programs, but this sustained effort is truly worthwhile!

The approach in presenting information is that of the voice of the advisor/counselor.

As you enter middle school, you will be meeting challenges in all areas of your life. Consider, for example:

- Academic Development.
- Extracurricular Activities.
- Sports.
- Community Activities/Service.
- Social Life.

How do you begin to prioritize and balance the demands in your life?

What we will try to accomplish in our group sessions is to make sure that you have some of the tools that can help you achieve success. Remember, this is your first priority—your primary job: to study and to achieve as much academic success as possible.

So, let's look first at who you are as a student. Complete Exercise 6.1 Who am I as a student? Ask yourself these questions and think about your responses. Heavy questions, but you probably have a good sense of your answers.

EXERCISE 6.1 Who am I as a Student?

Exercise 6.1 Who am I as a Student?	
What are my academic strengths?	
Where do I need to improve in my studies?	
How does this fit into knowing who I am and what I want to become?	

After students have completed Exercise 6.1, advisors/counselors may conduct a group discussion of student responses. Remember to keep digital and/or print copies of students' Exercise 6.1 for future reference.

Again, in the voice of the advisor/counselor….
Keep in mind your answers to the question: Who am I as a student? What we are going to do now is to talk about the **Study Skills and Habits** that can help you improve your academic performance.

First, let's review your **goals** as you entered middle school. While we may be focusing on the study skills and habits that will improve your academic performance, this approach may help you improve your performance in other areas of your life.

Take the time to review all your goals that we have discussed in earlier sessions. Redo Exercise 5.2.

Advisors/counselors may want to ask students if any significant changes have occurred in this exercise and the Why behind the change.

What Kind of Student Are You?

Now before we start talking about study skills, let's take a 10 question quiz that will help you to understand yourself as a student. *A note to the advisor/counselor: EDUCATIONPLANNER .org provides a 10 question survey that helps students to understand themselves as students. This assessment is available to middle school students.*[4]

Now, based upon your honest answers, you may need to change your approach to studying in class and at home. Even if you score a "You are doing well," remember that there is always room for improvement!

Before we talk about the **how** behind achieving your academic goals, let's make sure some words are part of your middle school vocabulary. You may know their definitions but think of these terms in a new way that is centered on you.

EXERCISE 5.2 Establishing Goals and How to Achieve Them

Goals	*How To Achieve*
Academic Achievement	
Extracurricular Activities	
Sports	
Other	

A YOU-CENTERED VOCABULARY

- Active Reader.
- Active Listener.
- Engaged Student.
- Proactive Student.

You will notice that the one word that is missing is **PASSIVE**. Passive and student do not belong in the same sentence!

An **Active Reader** notes:

- Bold headings and word sections in a textbook.
- Looks for the topic sentence in paragraphs and supporting ideas.
- Makes margin notes where possible.

In effect, **Active Readers** make the most use of their abilities when reading.

An **Active Listener**

- Pays attention throughout class.
- Takes notes as the teacher explains.
- Stays focused despite distractions.

Again, **Active Listeners** make the most of the classroom session.

An **Engaged Student**

- Participates in class discussions.
- Listens to what others have to say.
- Is truly involved in the subject work.

Engaged Students learn the most because they are genuinely involved in the learning process.

Finally, the **Proactive Student**

- Has clearly defined goals.
- Has developed strategies to achieve success.
- Is an active participant in the learning process.

Proactive Students assume responsibility for their own success.

Now let's explore learning styles.

A note to the advisor/counselor: Current research exploring learning styles suggests that the

idea that a student has one "fixed" learning style goes against the idea that learning should be flexible and

> adaptable. Mayer (2004) and Willingham (2005) emphasize the importance of varied instructional approaches and the ability of individuals to adapt their learning strategies in response to task demands and contextual factors... Proposed learning styles do not always fit disciplinary norms; for example, writing courses benefit from a significant verbal component, geometry courses from a visual component, and lab classes from an experiential component.[5]

Here, our emphasis is on the student's ability to develop learning strategies that adapt to different subject matter and tasks. At the same time, we are assisting them in identifying what they believe is their preferred learning style; in other words, how do they believe that they learn most naturally. With this awareness, students will more easily develop learning strategies as they adapt to classes that do not teach to what they regard as their preferred learning style. EDUCATIONPLANNER.org provides a 20-question survey that helps students to understand their preferred learning style. This assessment is available to middle school students and may be useful in explaining this topic to sixth graders. (Presented below.)

What's Your Learning Style?

Have you ever wondered why you do better in some classes than others? It may depend on your individual learning style. Your learning style influences the way you understand information and solve problems. At the same time, your preferred learning style may naturally adjust to the demands of particular subject areas. For instance, you may have a visual learning style in English class and become more of a tactile learner in science class and labs. Understanding the different learning styles may help you adjust more easily to the variety of classroom teaching techniques and subject matter that you will encounter and help you develop learning strategies that will help you learn more easily.

There are three primary learning styles:

- Auditory.
- Visual.
- Tactile.

Let's now take a 20-question survey that will help you to understand your preferred learning style.[6]

After students have completed the EDUCATIONPLANNER.org Learning Style Inventory, they will see what their preferred learning style is.

So, let's now look at the three major learning styles.

Your Preferred Learning Style

A note to the advisor/counselor: The following explanation of the three learning styles also poses questions for students who were unable to take the EDUCATIONPLANNER.org assessment.

Remember, a learning style is a way of learning. **YOUR preferred learning style is the way in which YOU learn best.** Three learning styles that are often identified in students are the Auditory Learning Style, the Visual Learning Style, and the Tactile/Kinesthetic Learning Style.

Are You an Auditory Learner?

Auditory learners learn best when they hear information. Do you seem to learn best in classes that emphasize teacher presentations and class discussions? Does listening to podcasts help you learn better? Do you find yourself reading aloud or talking things out to gain a better understanding? If YES, you are probably an auditory learner.

Are You a Visual Learner?

Visual learners learn best when information is presented in a written language or in another visual format such as pictures or diagrams. Do you do best in classes in which teachers do a lot of writing on the SMARTboard/whiteboard, provide clear handouts, and have students watch educational videos? Do you try to remember information by creating pictures in your mind? Do you take detailed written notes from your textbooks and in class? If YES, you are probably a Visual Learner.

Are You a Tactile Learner?

Tactile/Kinesthetic Learners learn best in hands-on learning settings in which they can physically manipulate something in order to learn about it. Do you learn best when you can move about and handle things? Do you do well in classes in which there is a lab component? Do you learn better when you have an actual object in your hands rather than a picture of the object or a verbal or written description of it? If YES, you are probably a Tactile/Kinesthetic Learner.

Your learning style is your strength. Go with it whenever you can. Later in middle school and in high school, you will have the opportunity to choose some of your classes. Select those classes that draw heaviest on your learning style. When you can pick a teacher, try to choose one whose teaching methods best match your learning style. When you decide upon a major and future career, keep your learning style firmly in your mind.

However, not all subjects or teaching methods will fit perfectly with your preferred learning style. How can YOU best learn to approach some subject areas in a new way that helps you learn more easily? Adaptability is the key word here. You may find that your learning style needs to vary in different subjects. For example, a visual learner may need to become more of a tactile learner in a science class. Now that you are aware of learning styles, develop learning strategies that will help you succeed in subject areas where you may need to learn in a way that is not your preferred learning style. Some strategies may require more planning and time for study and review, but you can succeed in these subject areas as well. Teachers and advisors are there to help you. Seek out after-school sessions in particular subjects for additional help when you feel that your study approach is not working.

Now, let's review **Executive Functioning**. In an earlier session, we talked about executive functioning and how important it is in the effective completion of assignments. All of you have done **Exercise 5.4 Executive Functioning** which has given you an idea of your **ExF Factor**—in other words, how well do you go about completing assignments. If you have any difficulty getting your work done and handing it in on time, now is the time to stop and look at ways to improve your executive functioning.

Four Keys to improving your **ExF Factor:**

- Effective Planning.
- Organization.
- Time Management.
- Avoiding Distractions.

So, let's talk about these **Four Keys** and how you can use them to achieve greater academic success.

A note to the advisor/counselor: As many of your schools use an LMS, particularly Google Classroom and Schoology, references in this section will be for both digital and print formats.

CHART 6.1 HOW GOOGLE CLASSROOM HELPS YOU WITH EXECUTIVE FUNCTIONING

Organization and Access:

- **Streamlined Class Management:** Students can easily access all their classes, assignments, and materials in one place.
- **Paperless Learning:** Classroom is designed to be paperless, with assignments and assessments stored digitally in Google Drive.
- **To-Do List:** Students can see a clear list of upcoming assignments and tasks.
- **Calendar Integration:** Automated calendar reminders help students stay on track with deadlines.

- **Offline Access:** Students can start work offline, review assignments, open Drive attachments, and write assignments in Google Docs, even without an internet connection.

Assignment Submission and Tracking:

- Students can easily submit assignments and track their progress.

Classwork Page:

- The Classwork page allows students to turn in assignments, answer questions, work with assigned documents, and check their work with rubrics.

Feedback and Grading:

- Teachers can provide feedback and grades directly within Google Classroom, and students can view rubrics alongside their work.

Read Along in Classroom

- This page has been upgraded to include more resources and educator insights, and new Google Classroom add-ons, new handwriting fonts for younger learners and a library of ready-to-use Google Slides templates to reduce the time spent on design.[7]

CHART 6.2 HOW SCHOOLOGY HELPS YOU WITH EXECUTIVE FUNCTIONING

- **Access to Learning Materials:** Students can access various learning resources, including lessons, videos, and documents, all in one place.
- **Assignment Submission and Tracking:** Students can submit assignments, track their progress, and receive feedback from teachers.
- **Gradebook and Progress Reports:** Students can view their grades and progress reports, allowing them to monitor their performance.

- **Communication and Collaboration:** Schoology facilitates communication and collaboration through discussions, forums, and messaging, fostering a sense of community.
- **Mobile App Access:** Students can access Schoology through a mobile app, allowing them to stay connected and engaged on the go.
- **To-Do List:** Students can see a list of upcoming assignments and deadlines.
- **Aggregated calendar:** Students can see all of their assignments and events in one place.[8]

First Key to Improve Your ExF Factor for Academic Achievement: Planning.

Let's Get Started: Using a Planner and/ or the To Do List and Calendar

Students may use a hard copy planner or access digital tools such as Calendars, To Do List and Gradebook and Progress Reports among other features on various LMSs. Some students may prefer to do a combination of digital and printed formats. What is essential here is that students develop a consistent method of using their planner daily, both digital and hard copy, to know what their assignments are and appropriate due dates. The following presentation may be adapted to both print or digital formats. Remember, even students who function in a LMS may also want to use a hard copy planner.

Why a Planner or a To Do List?

Both a planner and the To Do List will help you

- ◆ See what your assignments are.
- ◆ Organize your time more effectively.
- ◆ Schedule your time for long-term assignments.

- Keep track of your grades for each quarter.
- Include personal, extracurricular, and family events in your time schedule.

How to Make Effective Use of Your Hard-Copy Planner?

Some of the best hard copy student planners available do not list the month and the dates because it gives students more flexibility in incorporating school activities into their planning. This presentation is adapted to varying print and the appropriate LMS features.

A note to the advisor/counselor: Some sixth graders (and their parents) may initially prefer a hard copy planner; however, as students adapt to the LMS, they rely more on the digital planning features.

1. If you are using a printed planner, fill in the subjects on the left side of your planner. Do this for the first four weeks of school. Be sure that you keep the subjects in the same order for every week.
2. If you have a blank printed planner, fill in the month and date at the top of the planner page. Do this for at least the first four weeks of school. Make sure that you include the weekend dates—you will have homework over the weekend. Also note any days off and mark them in your planner. Continue to fill in months and dates as school progresses.
3. At the bottom of the weekly planner sheet, write in your own activities, such as extracurricular activities, sports, and family events. Remember you are developing **time management skills** as you plan for your homework and other commitments.
4. Write in assignments daily as needed. If you are assigned a long-term project or paper, write in the due dates now. You may also want to make a midway reminder note to help keep you on track to complete a long-term assignment.
5. Common features of LMSs are **To-Do Lists** and **Calendars.** As your teachers post assignments on their **Course Page,** these assignments are automatically added to the Calendar. These assignments also appear on the **To-Do-List**. You must

also use their **Calendar** to plan when to study for exams and complete assignments. Remember, it is essential even if you are using a hard copy planner, that you check the **To-Do-List** and the **Calendar** daily. You do not want to miss any assignments or to be late in handing in work!

You have a life outside of school and homework. Include social and family time in your printed planner so that you can effectively allot your time. If you are using digital planning, you should also include important personal events in your Calendar.

Now go to the back of the printed planner where you will find pages designed to help you keep track of your grades. Fill in your subjects and post any grades. However, as you become adjusted to your LMS, you may find it much easier to look at your grades digitally. Go to **Feedback and Grading** (Google Classroom) or **Gradebook and Progress Reports** (Schoology) where your teachers provide feedback and grades. (Other LMSs may provide similar options.)

Many middle schools have scheduled afterschool help in subject areas. Take advantage of extra help as needed! Ask your teacher for help before you receive a low grade on your report card.

The Second Key to Improve Your ExF Factor for Academic Achievement—Getting Organized and Staying Organized

Organization is a vital key to academic success, so let's talk about how you organize your schoolwork.

Many of your teachers will have you do work on your laptop, and you may even take notes that way, but there will always be hard copy materials that you should save. So now let's look at how to keep all of your subjects organized!

Different subject areas demand different kinds of organization, but the common theme is that you should have the necessary materials at your fingertips.

Individual subject teachers may tell you how they want your work organized. For example, your math teacher may want all of your work in one notebook.

If your teacher allows you to decide how to organize, try this method:

Materials needed: 2″ binders
Dividers with pockets
Loose leaf

Now Let's Organize your English and Social Studies Binder

First, label a divider English.

Then label dividers—class notes, homework, tests and quizzes, and handouts.

Place loose leaf in appropriate dividers.

Now label a divider Social Studies.

Then label dividers—class notes, homework, tests and quizzes, and handouts.

Adapt this system to every class. For example, in science class, you may need a divider for labs. If you follow this method of organizing your materials, you should be able to find whatever you need quickly.

At the end of a unit or a quarter, remove materials from your binder and place them in large envelopes or file folders for future use. Make sure to label! For example, Social Studies First Quarter or by topic—Roman Empire. This way, you know where your class notes, homework, and quizzes are if you need to study for a major semester exam.

Even if you use a LMS such as Google Classroom or Schoology, you may use binders for notes. While most assignments are submitted digitally, some teachers may ask for assignments on paper.

Clean out your binders regularly. You will be amazed at how many unnecessary papers you will find. They will only get in the way of your effective studying.

Let's look at the Third and Fourth Keys to Improve Your ExF Factor for Academic Achievement: Time Management and Avoiding Distractions.

Doing That Nightly Homework

You know that homework varies from night to night, but there is always something to do. Essential to homework completion and studying is to find a place in your home where you can work without distraction. And, of course, turn your phone off or put it on do not disturb. Distractions such as cell phones, whether scrolling or texting, can prevent you from becoming a successful student. Also, no visiting social media sites! This is your designated time for homework and study. You must avoid these distractions if you want to achieve real academic success. Again, find a quiet place in your home where you can study undisturbed! If this is not possible, perhaps studying at your local library for an hour or two may be a possibility!

A strong warning about the constant visiting of social media sites and doomscrolling: You may be harming yourself! Studies have shown that doomscrolling and excessive social media visits can reduce your attention span and diminish your ability to think critically. This is a serious warning![9]

Another warning—write your own paragraphs and essays. **Chat GPT** may be appealing, but you are now developing the skills for high school and beyond. Learn to write by doing it! Think of it this way: if you love basketball, how do you improve your shooting? Practice shooting, right? The same is true with writing skills. You will improve with practice and guidance from your teachers.

*Excessive visits to social media sites and doomscrolling are also presented in Chapter 5. These topics as well as the use of ChatGPT are addressed in the agendas of Parent Meetings in Chapter 14.

But before we begin, think about:

- What does it mean to **prioritize** and how can it help you?
- What does it mean to **review** and why is it important?
- What is the difference between doing **homework and studying**?

To prioritize is to decide which assignments are the most demanding and important to you. Look over your assignments and decide the order in which you will do your homework.

For some suggestions about completing your daily homework, complete the following chart to help you plan how you will approach your assignments. While it is important to decide the order of your work, you should also give yourself some short breaks (10–15 minutes). Also, don't forget that you have to eat dinner. Schedule that in as well. What is most important? A test, a paper? What will take the longest amount of time? Why?

Now plan your evening work. Look at all of your subjects. Keep track of how long you think that it will take to complete an assignment and how long it actually takes.

EXERCISE 6.2 Prioritize Your Work

	Review Class Notes	*Complete Assignment*	*Estimated Time*	*Actual Time*
5:00 pm				
5:30 pm				
6:00 pm				
6:30 pm				
7:00 pm				
7:30 pm				
8:00 pm				
8:30 pm				
9:00 pm				
9:30 pm				

Again, remember to turn your phone off or put it on do not disturb when doing your homework. You can use your breaks to catch up with your friends, but regard the time you have slated for your schoolwork as solid—no interruptions!

Why Is It Essential to Review?

Note that Exercise 6.2 has a column "Review Class Notes"—why is this important? Think of your mind as a computer. If you work on an essay on your computer and then forget to save it, you may lose it. The same is true for your memory. We all have short-term memory and long-term memory. Unlike saving on a computer, getting information into your long-term memory is not so easy, but study techniques that you incorporate into your homework can help you remember information.

- **To review** means to go over. As you do your homework each evening, spend a few minutes going over your class notes from that day. Believe it or not, this quick review can have a great impact.
- Also, once a week, **review** your class notes in each subject—this can also help in improving test grades.
- Use a highlighter to mark key ideas in your class notes when you **review.** Don't over-highlight, just mark essentials. This method of review helps to make ideas stand out in your mind.

The Difference between Homework and Studying

Doing your homework, your daily assignments, is important, but this daily work is the minimum. Studying demands review and understanding of the subject matter as thoroughly as possible.

When you review on a regular basis, what you are actually doing is a key step in studying for tests! So, when you are preparing for an exam, your review of important notes will already be in your long-term memory, helping you to achieve your academic goals!

How to Take Good Homework Notes

Doing your homework is the best way to be prepared for class. A common assignment, especially in Social Studies and Science, is to read and take notes on a number of pages in your textbook or from an online resource. Whether the source is hard copy or digital, you want to approach your homework in a way that will help you remember what you have read. Taking good notes is a great preparation and will ensure that you get the most out of class the next day.

Preview the material that you are to read. Remember that most textbooks use bolded and larger fonts to show subdivisions in the material that you will be reading. Your online resources may also do this. Look over the assigned pages, noting any **bolded subdivisions** and any individual words that are bolded. This tells you that these are key terms you will need to know.

Now it's time to read. Have questions in mind as you examine the material: Think about the **5Ws—the who, what, where, when, and why**—of the passage. It will help you to understand what you are reading.

Remember:

- Most paragraphs begin with **topic sentences** that express the meaning of the paragraph.
- Be aware of the **meaning of each topic sentence.**
- Look for the **supporting details** that explain the topic sentences
- **Take notes** as you read.
- Use the **bolded subdivisions** as a way of dividing your notes.

You do not need to do a formal outline—instead use spacing and indentation in your note taking.

Sample Notes on the Structure of Ancient Egyptian Society

The society of ancient Egypt was arranged like a pyramid.
Six Classes Existed.

- Pharoah.
- Government officials.
- Nobles and priests.
- Scribes and soldiers.
- Artisans and merchants.
- Peasants and slaves.

Taking notes this way shows the **main idea** of the structure of ancient Egyptian society, and the **indenting** lists the specific points.

Don't crowd your note taking; give yourself plenty of space. This makes it easier to review and study for exams.

Remember always to:

Preview
Note bolded subdivisions and bolded terms
Keep the 5Ws in mind as you read
Take notes using space and indentation

While we have been talking about note taking in Social Studies, you can adapt this method to your other subjects.

Taking Notes in Class

A rule to live by: Taking notes in class is always easier if you have done the homework. This helps you to know what the teacher is going to discuss and greatly enhances your understanding of the materials.

- Have your **homework** on your desk so that you know the order of the material.
- **Stay focused** despite distractions.

- Listen for **key phrases** such as "The key significance" or "The most important." This signals that these are ideas you want in your notes.
- Be aware of what the teacher writes on **the SMARTboard/ Whiteboard**—another signal for important material.
- **If you missed an important point,** note it and make sure to check with a friend after class so that you are not missing any information.
- Before beginning the next day's homework, make sure that your **notes are complete**.

Preparing for Tests—No Matter What Size

Taking tests is a fact of academic life. Now comes **The Big IF**—If you have done nightly and weekly review, preparing for an exam will be much easier for you. If you haven't, don't panic—you still have time to prepare.

Remember, we talked earlier about how to get information into your long-term memory? **Effective review** is the answer.

How to prepare for that unit test, midterm, or final exam?

- First, know **what material** will be on the test.
- **Organize** your notes, old tests, quizzes, and relevant handouts.
- Establish **a realistic time schedule** for studying for your big test or exam. Do not leave all of your studying until the night before the test.
- Start **to review materials in realistic amounts**—for example, a chapter at a time for a midterm.
- Use a **highlighter** to mark key ideas.
- **Box** in key terms, names, places, events with a red pen.

Doing this method of review will help ideas stand out when you take a test.

Make a **cover sheet** for each chapter or unit. For example, for a test on ancient Egypt, use a chart to mark key people, places, events, and terms. **Do not identify—simply list.** After you have reviewed your notes and listed items, use the cover sheet to quiz yourself.

EXERCISE 6.3 Making Cover Study Sheets for Unit Exams

Sample: Ancient Egypt Cover Sheet

Key People	*Places*	*Events*	*Terms*

Make sure to keep **cover sheets** with the appropriate materials. These sheets can be invaluable in studying for major exams such as unit tests, midterms, and finals.

While the cover sheet example is for a Social Studies test, you can adapt this method to other subjects. For example, list short stories, authors, themes, characters, etc. for English.

For some subjects, it is helpful to make flashcards to quiz yourself. For example, you may want to use index cards to help you master foreign language vocabulary. On each card, print the word on the front and definitions on the back. Make sure to keep cards in a labeled envelope. For example, Spanish Chapter 2 vocabulary. Store these at home or in a pocket folder in your binder. You may also want to use flashcards for other subjects. If you want to make flashcards digitally, **Quizlet** is a helpful online source for you. Here you make digital flashcards to quiz yourself on multiple subjects.[10]

Review, review, review is what studying is all about!

Taking the Test

Some Dos and Don'ts About Test Taking

Definitely Do

- Get a good night's sleep and eat breakfast.
- Be conscious of time allotted for the test.

- Look over the entire exam.
- Read directions. It is important that you know what you are being asked to do. If you are uncertain, ask for clarification.
- Work systematically. Do not jump around answering questions; you may end up leaving some questions blank.
- Be aware of the number of points (credit) attached to a question. Don't spend so much time on one question that you are unable to answer others.
- Stay focused during the entire exam.

Definitely Don't

- Stay up all night trying to cram.
- Skip reading directions because you think it will take too much time.
- Answer some questions and not others, thinking that you will be able to come back to them later. This usually ends up in blank answers and misplaced answers on the answer sheet.
- Spend too much time on one essay question or math problem so that you are unable to finish other questions.
- Rush through the exam thinking that you will have time to check your answers.
- Finish early and relax rather than checking your answers.

Remember, you are developing who you are as a student. Now is the time to master the study skills and habits that not only will help you achieve your goals in middle school, but also assist you in high school and beyond! Remember, you are on a K–12 journey and these middle school years are an opportunity to grow and think about the possibilities that your future may hold! As we move ahead, we will venture into career exploration—a whole new world of possibilities awaits!

A Final Note

Truly understanding executive functioning and the ways to improve a student's own ExF factor is an ongoing process that for some students may involve an entirely new approach to homework and study. While these students may need more attention and encouragement, they will learn what may be a life-altering approach to their studies, one that will truly provide the foundational basis of study skills and habits that are so necessary for success. As educators, we have both encountered students struggling to acquire better study skills and habits in high school. These experiences have only reinforced our belief that middle school is the crucial time period to introduce the concept of executive functioning and its related components to students.

Notes

1 Lemberger-Truelove, Matthew et al. "Growth in Middle School Student's Curiosity, Executive Functioning and Academic Achievement." *Professional School Counseling*, May 2021, Vol. 24, No. 1b, Special Issue: Educator–Counselor: A Nondual Identity for School Counselors, p. 1.
2 "The Best Way to Study in Jr. High: Valuable Skills." *K12.* December 2024. https://www.k12.com/study-resources/best-way-to-study-in-jr-high/
3 "College- and Career-Readiness Standards." *Teachers College Columbia University.* https://sri.tc.columbia.edu/our-framework/college-and-career-readiness-standards/
4 "Self Assessments- Find Out More About Yourself." *EducationPlanner.Org.* https://www.educationplanner.org/students/self-assessments
5 "Learning Styles as a Myth." *Poorvu Center for Teaching and Learning.* https://poorvucenter.yale.edu/teaching/teaching-resource-library/learning-styles-as-a-myth
6 "What's Your Learning Style?" *EducationPlanner.Org.* https://www.educationplanner.org/students/self-assessments/learning-styles
7 "Google Classroom Features." *Google for Education.* https://edu.google.com/workspace-for-education/products/classroom/editions/
8 "Schoology Learning." *PowerSchool.* https://www.powerschool.com/solutions/personalized-learning/schoology-learning/
9 "Doomscrolling and Brain Rot: What Are They?" https://www.nationwidechildrens.org/family-resources-education/700childrens/2025/01/doomscrolling-and-brain-rot
10 Quizlet. https://quizlet.com/

7

Using Technology in Middle School

In this chapter we present thinking about how students with the help of advisors/counselors can use innovative technologies. While most of the relevant technologies will be addressed in the individual chapters, this chapter provides an overview of important "big picture" ideas that are crucial to understanding how middle school students use technology. Specifically, advisors/counselors need to consider two overarching changes: in exposure to media and in our continuing understanding of the learning process in the digital age. The challenge is not about how middle school students interact with technology, but rather how advisors/counselors and students continue to keep a balance between making schoolwork exciting through technology, yet maintaining needed rigor in critical thinking, reading, analysis, and synthesis development.

Middle School in the Post-COVID Era

In the post-COVID period, the role of technology in the classroom has increased dramatically. For many school districts, the pandemic hastened the introduction of school-issued laptops

DOI: 10.4324/9781003597292-9

and online learning management systems such as Schoology and Google Classroom. At the same time, the COVID period dramatically contrasted school districts as many urban and rural areas struggled with paper assignment packets, uncertainty about whether computers were in the home as well as poor internet access, particularly in mountainous areas of the country. As the years of the pandemic progressed, more school districts were able to engage in remote learning through school-issued laptops and Chromebooks. As school districts gradually reopened, the new era of learning management systems became a reality in American education.

A recent report from the Center for American Progress stresses that access to high-quality educational technology improves student learning and preparedness for postsecondary success. Moreover, middle school students seem to be interested in technologies that are relevant. Students must have digital literacy: the ability to use digital tools and technologies to learn and interact. Teachers should accent how essential effective integration of technology is:

> When integrated effectively, technology can increase student engagement, improve collaboration, and promote choice in student-centered learning, …while giving students practice with basic digital skills and exposure to more advanced skills…. Digital literacy is becoming increasingly important to successfully preparing students for both college and careers; these technological advancements are creating new job types that the workforce must be prepared to fill.

The need for digital literacy is an ongoing concern for all students as the country struggles with the growing digital disparities nationwide.[1] Most middle school students are already familiar with technology in their everyday lives. For so many young adolescents, the integration of digital learning is a welcome aspect of learning in the post Covid classroom.

However, American education is now grappling with the magnitude of the widening digital learning gap. While poorer school districts strive to lessen the digital divide, the task is formidable. Sean Reardon, a Stanford sociologist, notes that educational achievement differences are more closely aligned with family income than any other demographic factor. Researchers have also shown that these achievement gaps have already begun growing by the time lower-income children reach kindergarten.[2] In addition to economic factors, most children with technology and internet access are White or Asian American while many Black and Latinx may lack technological resources. These findings are particularly troubling, especially as schools nationwide not only move into the new era of digital learning, but also continue to struggle with the effects of school closures during the COVID period, especially in areas where remote learning presented multiple challenges.

Another worrisome component is location. "When household income is held constant, students who live in rural areas are less likely to have internet access than students who live in urban areas."[3] Students and educators in many areas of the United States struggle with inconsistent internet connection. A survey by Project Tomorrow found that, "Nationwide, more than half of middle school and high school students (53 percent) reported slow or inconsistent Internet connectivity at school and identified that problem as the main obstacle for using instructional technology more effectively."[4] Clearly, lack of access to the internet impacts the education of students, presenting a grave issue of concern as the digital divide continues.

Under President Joe Biden, the Infrastructure Investment and Jobs Act of 2021 has provided a historic $65 billion investment to expand internet access and use throughout the United States. By November 2023, all 50 states and 6 territories had submitted their Broadband Equity, Access and Deployment Program Initial Proposals.[5] A prime example is West Virginia. Currently Through Internet for All, West Virginia, the WV Office of Broadband, WV Broadband Enhancement Council, and partners are joining with residents, businesses, nonprofit organizations, and numerous partners to develop West Virginia's Digital Equity Plan and Broadband Equity, Access, and Deployment (BEAD) Five-Year Action Plan as directed by the National Telecommunications and

Information Administration. The Internet for All West Virginia initiative is designed to achieve digital equity throughout the Mountain State.[6] Currently in mountainous areas where internet was sparse at best, school districts and students will now have greater access to digital learning. In addition, as part of the American Rescue Plan of 2021, the Elementary and Secondary School Emergency Relief Act, also known as ESSER, provides federal funds to help schools recover from the COVID pandemic period. Districts across the country have used ESSER funding to help create more collaborative classrooms by investing in the latest technology to foster an interactive learning environment for their students. Currently, ESSER funding has been extended to March 2026, but districts must have allocated funds by September of 2024.

American education continues to face profound challenges in the pursuit of fair educational opportunities for all. In January of 2024, the United States Department of Education released the 2024 National Educational Technology Plan (NETP): A Call to Action for Closing the Digital Access, Design and Use Divides. "From the American Rescue Plan to the Bipartisan Infrastructure Law and more, the Biden-Harris Administration has made bold investments aimed at closing the digital divide and ensuring all students can equitably access the latest digital tools and technology," said US Secretary of Education Miguel Cardona. "As we work to Raise the Bar in education, it's essential we focus on empowering teachers to become designers of active learning, by using technology in effective ways to engage and inspire students." The 2024 NETP is a forward-thinking approach to reframing and realizing the potential of educational technology to enhance the instructional core, reduce achievement gaps, and improve student learning in our schools.[7]

The 2024 NETP presents Three Divides:

1. **Digital Use Divide**: Inequitable implementation of instructional tasks supported by technology. On one side of this divide are students who are asked to actively use technology in their learning to analyze, build, produce, and create using digital tools, and, on the other, students encountering instructional tasks where they are asked to use technology

for passive assignment completion. While this divide maps to the student corner of the instructional core, it also includes the instructional tasks drawing on content and designed by teachers.
2. **Digital Design Divide:** Inequitable access to time and support of professional learning for all teachers, educators, and practitioners to build their professional capacity to design learning experiences for all students using edtech. This divide maps to the teacher corner of the instructional core.
3. **Digital Access Divide:** Inequitable access to connectivity, devices, and digital content. Mapping to the content corner of the instructional core, the digital access divide also includes equitable accessibility and access to instruction in digital health, safety, and citizenship skills.

The NETP maps each of these divides to the instructional core (students, teachers, and content) focusing on how schools, districts, and states can use educational technology to help design learning experiences that improve student access to educational opportunities and their outcomes. The NETP provides action-oriented recommendations for states, districts, and school leaders alongside examples from all 50 states, the District of Columbia, and several territories for addressing disparities in educational technology and closing the three divides. To assist in the funding of the NETP's recommendations, states and districts can utilize funds from existing federal grants, including those under the Every Student Succeeds Act (ESSA), the Individuals with Disabilities Act (IDEA), and the Infrastructure Investment and Jobs Act (IIJA). The Biden administration has structured a path to assist states and districts in a strong effort to narrow the digital divide existing in the country.[8]

We live in a world of rapidly advancing technology. It is essential for middle schoolers to understand technology as it increasingly impacts global society and the economy.

> At best, students should be exposed to various types of technology and understand how they are used in the workforce. At a minimum, students should be

comfortable with creating and saving documents, researching credible sources, and other basic computer skills. This knowledge is a critical component of college and career readiness.[9]

Teachers, counselors, and administrators are actively working to assist middle schoolers. The number of classrooms using the latest education technologies is growing. A recent study by the University of Kansas reports a widespread adoption of technology in schools. By 2024, a considerable 80% of global schools had implemented technology for educational use. Classrooms have various digital devices like computers, tablets, smart boards, and educational software and apps. These technologies have not only replaced traditional teaching tools but also significantly enhanced the learning environment.[10] For all educators—teachers, counselors, and administrators—enhanced digital literacy is essential professional development so that they may more effectively use technology to help students thrive in the digital learning environment of today's classroom. Just as school districts who were able to incorporate learning management systems (LMS) such as Google Classroom and Schoology in the fall of 2020 of the COVID pandemic, many educators spent significant time and energy learning these new systems during the previous summer. School districts should be aware of how well technology is integrated into the classroom and provide ongoing professional development as needed.

A Look at the New Middle School Classroom

The new middle school classroom differs dramatically from the days of blackboards and students writing in notebooks.

Some Prevalent Technologies needed in Modern Classrooms

1. **Laptops and tablets** are essential for facilitating interactive and portable learning environments.

2. **SMARTboards/interactive whiteboards** increase student engagement by incorporating multimedia resources and interactive activities.
3. **Learning Management Systems (LMS)** such as Google Classroom and Schoology are crucial for overseeing student information and enabling online education.

Crucial to effective digital learning is an appropriate LMS that is specifically devised for educational institutions. This LMS provides an integrated platform for teaching, learning, and school administration as well as improving parent–teacher communication gaps.

How LMS Tools Can Help Educators

- Share learning and review materials expeditiously.
- Create a safe virtual space to learn for all students.
- Communicate with parents.
- Manage student grades.
- Provide lesson feedback to groups and individuals.
- Poll students.
- Keep track of student learning and manage improvement.

Some Popular Learning Management Systems

- Blackboard is used by 28% of institutions and 37% of enrollments. Blackboard is known for its remarkable analytics and reporting features. This LMS is widely used by higher education institutes.
- Canvas accounts for 21% of institutions and 27% of enrollments. This LMS is most famous for providing tools that make the learning process easier and increase the learner engagement of online learners. Like Blackboard, Canvas is widely used in higher education institutions.[11]
- Moodle as of late 2023 had 16% of the North American market share. Similar to Blackboard and Canvas, this

LMS is widely used in higher education. Moodle has maintained its popularity by keeping it simple for users.[12]

Two LMS Platforms Widely Used in K–12 Education

- Google Classroom is a cloud-based LMS that is **free** to use for educational institutions. It integrates with other Google products, such as Google Drive, Google Docs, and Google Forms, to make it easy for teachers to create and manage assignments, grade student work, and collaborate with students.[13]

  Google Classroom also facilitates creating virtual classrooms and connecting to students in real time. This LMS is compatible with all devices including computers, tablets, and mobile phones. Google Classroom has 100 MB of storage with no limit to the number of users that an institution may add. Increased storage limit and access to additional features is available for a fee.[14]

  Google Classroom is not commonly regarded as a complete LMS despite its many capabilities for class organization. However, it offers numerous advantages similar to those found in full-fledged LMS platforms, making it a highly effective tool for classroom management and organization. A major appeal of this system is that it is free, which may be a key consideration for many school districts.
- Schoology is a product designed as an academic LMS specifically for K–12 schools. Like Moodle, Schoology is often high on the list of the best LMS software because it connects the people, content, and systems that fuel education. Its additional integrations include collaborative working with Microsoft products, Google applications, and PowerSchool. This LMS charges a fee.

  Schoology offers a variety of assessment methods including grades, exams, and homework. This LMS can be integrated with the school's current grading system. Also included in Schoology are timed quizzes and a feature that can be used to send daily reminders to students. Additionally, this LMS enables teachers to upload Google docs as assignments as well as the sharing of learning materials. Additional features include integration of the report card system and attendance records as well as parental portals to view their children's learning process.

A recent study found that Schoology is the LMS that aids the learning and training process for all K–12 classrooms by providing different options of digital content to active users and saving time through enhanced collaboration between teachers. Schoology works to enhance teacher effectiveness, facilitate data-driven decisions that the learners can benefit from, save valuable time for teachers, and provide engaging classroom environments.[15]

A Final Note About Schoology

The top three geographies of Schoology for learning-management-systems are the United States with 1,552 (87.98%), Canada with 53 (3.00%), United Kingdom with 49 (2.78%) customers respectively.[16]

The new age of LMS has dramatically changed how students learn, how they do homework, and how they submit assignments. Teachers reach out to both students and parents through these system portals, making it less likely for students to fall through the proverbial cracks. Students across all learning levels are using technology in class and at home, not only to develop enhanced critical thinking skills, but also to achieve and expand digital literacy.[17] Helping students to maximize their understanding and use of their LMS is a perfect subject matter for advisory groups and group counseling sessions. As referenced below, once acclimated to the features of the LMS, students may use programs such as the To Do List and the Calendar for more effective time management and completion of assignments. Easy communication between students, parents, and teachers is also available through portals.

How Learning Managing Systems Help Middle School Students and Parents

What is important for students and parents to understand is that a LMS provides tools that are digital; these tools do not do homework or studying for the student. Rather these digital tools enhance key components of executive functioning such as the ability to organize, plan, and manage time so that assignments are submitted on due dates. Both Google Classroom and Schoology provide ways for students to develop their executive functioning abilities that are so essential for personal and academic success.

- Google Classroom's To Do List and Calendar Integration and Schoology's To Do List and Aggregated Calendar are critical tools for developing effective time management and planning strengths. While some students may also choose to use a traditional planner, what is being developed here are the essentials needed for a strong fundamental basis of enhanced study skills and habits as discussed in Chapter 5. Additionally, both systems have Assignment Submission and Tracking features so that students may easily see if they are missing any assignments. These features of both Google Classroom and Schoology are the new digital tools necessary for young middle school students to use as they develop their executive functioning abilities, enhancing their ExF factor that is at the basis of strong study skills and habits.
- Another key component of both LMSs is the organization of learning materials. Google Classroom features a Streamlined Class Management feature where students may access all their classes, assignments, and materials in one place. Additionally, students may use the Offline Access feature where they start work offline, review assignments, open Drive attachments, and write assignments in Google Docs, even without an internet connection. While most students today do have internet access,

this feature may help students where internet connections may be intermittent or inadequate. Similarly, Schoology provides an Access to Learning Materials feature where students can access various learning resources, including lessons, videos, and documents, all in one place.

Again, while these organizational features provide excellent assistance in organizing materials, students may use them in conjunction with traditional organizational methods. Students will have paper notes, handouts, essays, and even at times, homework that is submitted using the traditional hard copy method; these materials should be effectively organized. Notebooks and binders still exist in this new digital learning space alongside textbooks, hard copy novels, and short stories. Particularly as students move on to high school, academic demands will increase, especially in honors and advanced placement classes where students need to review materials in large amounts for major tests. The AP United States History class is a prime example where students will use notes, both digital and hard copy, from September through mid-May to prepare for this exam. This topic is addressed in Chapter 6.

- A feature of both systems that parents may find very useful is the access to communicate with educators through the LMS. Google Classroom's Real-time Communication enables announcements, discussions, and short question and answer availability. Teachers can also send email summaries to parents/guardians. Likewise, Schoology's Communication & Collaboration platform includes messaging, discussion boards, and updates to connect teachers, students, and parents. It also supports group collaboration for teams or clubs. Again, these features promote communication between parents and teachers; however, this feature is not intended to replace the essential Parent Meetings that will truly help parents to understand the work that is being done in middle school. (See Chapter 14 for more information regarding Parent Meetings.)

Today's middle school differs dramatically from the pre-COVID era. However, the aim of establishing the foundational basis of the College and Career Readiness Standards remains. The methodology of using digital tools as well as traditional approaches helps to prepare young adolescents as they grow in their K–12 journeys. Middle school students are actively engaged in this learning process that will promote their development and prepare them for the next step of their journey—high school.

For more information about Google Classroom and Schoology, visit:

Google Classroom Features. Google for Education, https://edu.google.com/workspace-for-education/products/classroom/editions/

Schoology Learning. PowerSchool, https://www.powerschool.com/solutions/personalized-learning/schoology-learning/

Just as the recent pandemic exposed the Digital Divide in the United States, let us now look at two very different school districts and how each is moving to make greater educational opportunities more available for all students. Both the urban New York City Public School System and the rural Randolph County Public School System in West Virginia have found a way to provide greater digital learning for their students. In both instances, these districts used federal, state, and local funding.

In 2023, New York City purchased 550,000 iPads and 175,000 Chromebook laptops for about $360 million for students in NYC public schools. Every student in the system has a NYCPS account that allows access to multiple platforms including TeachHub, Google for Education, iLearnNYC, and Microsoft Office among others. Students who are issued a Department of Education-loaned iPad can keep it throughout their time enrolled in NYC public schools. New York City Public Schools use Google Classroom.[18]

A more rural example is West Virginia's Randolph County School District that has 14 public schools. These schools include elementary schools, middle schools, and high schools. In 2024, Randolph County supplied Chromebooks for all students (K–12 grades). Chromebooks are property of the district and are borrowed by students. In grades 3–12, students use their Chromebooks in class and take them home each night to complete assignments. Randolph County School District uses Google Classroom. One example of this district is Pickens School, nestled in the Appalachian Mountains of central West Virginia. The student population of this K–12 school is approximately 34 students. As of 2024, Pickens School has a fiber optic (high speed) connection to the Internet—no small feat in this mountainous area. Middle and high school students take some of their elective classes online through the WV Virtual School and math classes through a video connection to the much larger Elkins High School.[19]

The Trump Administration

President Donald Trump has advocated returning the responsibility of education to the state governments. Consequently, his administration has moved to dramatically downsize the United States Department of Education. According to the Center for American Progress (CAP), since Trump took office in January 2025, the US Department of Education's workforce has been dramatically reduced. About 58% of staff have received layoff notices or accepted resignation offers this year, but after multiple reductions in force and reinstatements, the total reduction by the Trump administration was approximately 48%. However, the Trump administration in March of 2025 fired another 465 staff members including staff in the Office of Elementary and Secondary Education (OESE), Office of Special Education and Rehabilitative Services (OSERS), and Office for Civil Rights (OCR). While this action is being challenged in court, the outcome is uncertain. "If the Trump administration ultimately moves forward with this reduction, the offices that support the education of roughly 50 million K-12 students by enforcing civil rights laws and overseeing funding distribution for Title I schools, special education, rural schools, and more will be left with little to no staff."[20]

Later in November of 2025, the Trump administration further continued the dismantling of the Department of Education by transferring responsibilities to other agencies.

> The Department of Education announced six interagency agreements (IAAs) …with the Department of Labor, the Department of Health and Human Services, the State Department and the Department of the Interior to co-manage or take a growing role in managing certain offices and programs.

As of January 9, 2026, Education Secretary Linda McMahon announced that during the year 2025, 39% of the staff of the Department of Education had been eliminated, reinforcing President Trump's belief that education should be the concern of the states and local communities.[21] Furthermore, on June 30, 2025, the Department of Education announced to state education associations that it would withhold almost $6.9 billion in funding approved by Congress. But on July 18, after facing bipartisan pressure, the administration said it would release $1.3 billion in grants for summer and after-school programs if states provided new assurances as to how the funds would be spent.

> Still, schools have to prepare to begin the next school year without the other $5.5 billion in previously approved funds that they were counting on. Districts in every state are scrambling to figure out how they will continue to meet student needs if the rest of the funding is not released.[22]

According to EducationWeek, in Trump's first year of his second term as president, at least $12 billion in school funding has been disrupted in 2025.[23] Consequently, school districts nationwide face severe budget cuts.

A Final Note

As American schools continue to grapple with multiple issues including the Digital Divide, we must remember how aid to less affluent districts is essential to the belief in equal opportunity that is so deeply threaded into American history and values.

National aid to American education dates back to the Land Grant Act of 1787, part of the Northwest Ordinance. A founding principle of America is the belief that citizens should be educated for democracy to succeed, and that equal opportunity is for all. As a nation, we must continue to aid districts in need and to fight to close the Digital Divide that impacts many citizens.

Notes

1 Shoemaker Demio, Paige. "How States and Districts Can Close the Digital Divide To Increase College and Career Readiness." *Center for American Progress,* July 2024. https://www.americanprogress.org/article/how-states-and-districts%20-can-close-the-digital-divide-to-increase-college-and-career-readiness/
2 Bruce, Alyvia. "Bridging the Technological Divide in Education." *Harvard Political Review,* November 2020. https://harvardpolitics.com/education-tech-gaps/
3 Bruce, Alyvia. "Bridging the Technological Divide in Education." *Harvard Political Review,* November 2020. https://harvardpolitics.com/education-tech-gaps/
4 "K-12 Digital Divide Persists in Device Access, Media Creation.". *GovTech,* May 2024. https://www.govtech.com/education/k-12/k-12-digital-divide-persists-in-device-access-media-creation
5 Broadband USA. https://broadbandusa.ntia.gov/
6 Internet for All West Virginians. https://internetforallwv.wv.gov/
7 "2024 National Education Technology Plan." *American Federation of School Administrators,* January 2024. https://www.theschoolleader.org/news/2024-national-education-technology-plan
8 *A Call to Action for Closing the Digital Access, Design, and Use Divides.* 2024 National Education Technology Plan, Office of Educational Technology, US Department of Education, January 2024.
9 Shoemaker Demio, Paige. "How States and Districts Can Close the Digital Divide To Increase College and Career Readiness." *Center for American Progress,* July 2024. https://www.americanprogress.org/article/how-states-and-districts%20-can-close-the-digital-divide-to-increase-college-and-career-readiness/
10 "How Teaching With Technology Has Transformed Education." *KU School of Education and Human Sciences,* September 2021. https://educationonline.ku.edu/community/how-teaching-with-technology-has-transformed-education
11 "What is the Most Commonly Used LMS?" *TopAdvisor,* February 2024. https://www.topadvisor.com/resources/what-is-the-most-commonly-used-lms-a11052#:~:text=Blackboard%20is%20the%20most%20popular,K%2D12%20and%20higher%20education
12 "Moodle Vs Skillsoft: In-Depth Comparison." *6Sense.* https://6sense.com/tech/learning-management-systems/moodle-vs-skillsoft#free-plan-signup
13 Google Classroom Community. https://support.google.com/edu/classroom/community?hl=en

14 "What is the Most Commonly Used LMS?" *TopAdvisor,* February 2024. https://www.topadvisor.com/resources/what-is-the-most-commonly-used-lms-a11052#:~:text=Blackboard%20is%20the%20most%20popular,K%2D12%20and%20higher%20education
15 "Which Is Better Google Classroom or Schoology?" *TopAdvisor,* July 2022. https://www.topadvisor.com/resources/which-is-better-google-classroom-or-schoology-a11060
16 "Schoology." *6Sense.* https://6sense.com/tech/learning-management-systems/schoology-market-share
17 "How Teaching With Technology Has Transformed Education." *KU School of Education and Human Sciences,* September 2021. https://educationonline.ku.edu/community/how-teaching-with-technology-has-transformed-education
18 "Google Classroom." *NYC Public Schools.* https://www.schools.nyc.gov/learning/digital-learning/applications-and-platforms/google-classroom
19 Pickens School. https://pickens.rand.k12.wv.us/o/pies/page/about
20 James, Weadé and Paige Shoemaker Demio. "The Trump Administration's Latest Staffing Cuts at the Department of Education Threaten Children's Success Across the Country." *Center for American Progress,* October 2025. https://www.americanprogress.org/article/the-trump-administrations-latest-staffing-cuts-at-the-department-of-education-threaten-childrens-success-across-the-country/
21 Colton, Emma. "Power Stripped from Education Department in Latest Trump Administration Move to Dismantle It." *Fox News,* November 2025. https://www.foxnews.com/politics/power-stripped-from-education-department-latest-trump-administration-move-dismantle
22 Litvinov, Amanda. "Trump's Unlawful Funding Freeze Threatens Critical Student Supports." *NEAToday,* July 2025. https://www.nea.org/nea-today/all-news-articles/trumps-unlawful-funding-freeze-threatens-critical-student-supports
23 Lieberman, Mark. "In Trump's First Year, At Least $12 Billion in School Funding Disruptions." *Education Week,* January 2026. https://www.edweek.org/policy-politics/in-trumps-first-year-at-least-12-billion-in-school-funding-disruptions/2026/01

8

Career Exploration in Middle School as a Motivational Technique

The focus of Chapters 8 and 9 are closely linked. In Chapter eight, middle school students begin an initial career exploration which flows naturally into Chapter nine where they initiate a preliminary college search. Helping middle schoolers broaden their perspectives about their lives in the postsecondary world is our aim in these chapters. Middle school students are developing physically, cognitively, and psychosocially; in other words, they are amid multiple changes. What we have done in Chapters 5 and 6 is to make them aware of executive functioning and its components and the how behind developing solid study skills and habits that are the foundation for high school and beyond. Now, as students begin an initial career search, not only will this exploration broaden their views of professions and occupations, but it will also provide motivation as they come to understand the education needed for future career paths.

DOI: 10.4324/9781003597292-10

Exploring Possible Careers

Middle school students are at a focal time in their lives to think about future aspirations. While they may be considering the age-old question: "What do you want to be when you are grow up?" given the digital age, their world is filled with visions of multiple career possibilities. Research supports the idea that adolescents are open to future possibilities. In 1986, Albert Bandura theorized that self-efficacy is determined by a person's judgment about how well they can complete a task.[1] Later, Lent, Brown, and Hackett adapted Bandura's work to their Social Cognitive Career Theory (SCCT; Lent, Brown, & Hackett, 1994), which suggests that career choice is influenced by outcome expectancies, career interests, and career self-efficacy. Most importantly, as noted in Chapter 2, the Gaertner and McClarity research project focused on promoting self-awareness in middle school children. In this research project, Gaertner and McClarity created a diversity of middle-school variables synthesized into six factors: achievement, behavior, motivation, social engagement, family circumstances, and school characteristics. Their research found that middle school factors explain 69% of the variance in college readiness, and most notably, in motivation and behavior.

> These results suggest a variety of factors beyond academic achievement, most notably behavior and motivation – factors that could indicate and that contribute strongly to being prepared for post-secondary study…. This approach aims to provide earlier, more actionable information for students, teachers, and parents.[2]

What this study affirms is that career and college readiness should be a definite part of the middle school experience. As middle school students explore both careers and colleges, they are looking at their possible future selves at a time in their young lives where this early research may have a highly motivating influence on them.

Curry and Milson concur that middle school students are at a stage of development where they are open to exploring future career possibilities. At this stage of their young lives, adolescents "are working to establish their identities, which involves being able to discuss their strengths, weaknesses, interests, values and desires for the future."[3] This suggests that adolescents are developing a strong sense of themselves and are open to exploring their future self-identities in career possibilities. Furthermore, as Curry and Milson note, college and career awareness in young adolescents is largely influenced by self-concept and an awareness of their interests, abilities, values, and aptitudes. "In short, knowing oneself is the key to early adolescent career growth."[4] As middle school students explore possible career endeavors, they are also examining their own self-concept and attitudes about what their future selves may attain. Furthermore, Curry and Milsom advise school counselors to use assessments to help middle school students understand the skills, aptitudes, interests, and values that they hold. Such activities "may be helpful for promoting self-exploration and moving adolescents from diffusion and foreclosure to moratorium (Marcia, 1987) where they can actively begin to question who they are in relationship to their future career and college choices."[5] A clear goal of this book is to assist young adolescents in this examination of what will hopefully result in a deepening sense of the future possibilities open to them. As we continue in this chapter, we provide multiple inventories that are designed to help middle school students to understand themselves in a broader fashion. As students explore these inventories, it is also of key importance that they understand the necessary education required for particular careers. As middle school students explore careers and colleges, advisors and counselors should ensure that all middle schoolers remain open to the possibilities available in higher education. (This topic is further explored in Chapter 9.)

Further research has shown that career exploration is a positive motivating factor for this age group. According to the Hechinger Report, advocates argue that exposing students to potential careers in middle school, rather than waiting until high school, gives them time to take the classes and extracurriculars that will get them to their goals—and the opportunity to change course while the stakes are still low.

Furthermore, "the Dallas Independent School District, the second-largest district in the nation's second-largest state, has long offered career exploration courses to its seventh and eighth graders. But this year it expanded one of the classes.... to every middle school in the district." Nancy Deutsch, a University of Virginia professor and director of Youth-Nex: The UVA Center to Promote Effective Youth Development concurs, observing that, "Early adolescence is such a huge time for identity development, when young people are asking, 'Who am I, and who do I want to be?'... Career exploration capitalizes on this innate drive, encouraging students to try on possible future selves."[6] Particularly in this age of a rapidly changing global economy, making students aware of career areas that interest them is so important in developing their future career possibilities. Middle school students are truly open to career exploration as they come to understand themselves not only as students but also as people with future aspirations.

As part of the shared agenda of the advisory/group counseling sessions, both early career and college exploration are main concerns as middle school students are eager to look at what life ahead may offer them. Particularly in this time of a rapidly evolving, digital workplace, adolescents not only need to examine future careers, but also, they may be excited by this early research. It is difficult to imagine an occupation that has not been dramatically transformed by the digital age. The modern workplace has changed, and middle school students will be able to hear professionals describe their work duties in this new age as they explore career sites in the advisory/group counseling sessions.

As we present curriculum ideas to promote career exploration, our aim is to help all middle school students look at the possibilities ahead of them. It is the perfect time to start on the journey of finding a fulfilling work realm. Again, the emphasis is on all middle school students in grades 6–8. Our curriculum suggestions are not geared to student performance on district or state-wide assessment tools or on academic performance. We believe that all middle school students will benefit greatly by exploring career interests through websites and activities designed to assist them as they discover various career paths. The career exploration that students do in this chapter flows naturally into Chapter 9 where students will begin initial research of colleges.

A side note: In our work with high school students, we as coauthors have seen how older students, sophomores and juniors, may bring multiple concerns to career exploration. Once in high school, students become increasingly aware of how school performance, particularly how their grade point average, GPA, may impact college admissions. Added to this is the high school students' increasing concerns about performance on standardized testing such as the ACT and the SAT. These concerns do not mean that career exploration should not occur in high school. However, the role of the advisor/counselor must address these issues, assuring students that multiple career paths exist for all students. What we have learned as educators is that career exploration in middle school lacks the tensions that high school students experience. Middle school students engage in this exploration with excitement and enthusiasm. Indeed, discovering careers that interest them and understanding the education required may be a highly motivating factor for middle school

Career exploration presented in Advisory Groups or Group Counseling sessions allows students to be involved in an ongoing pursuit of their career interests. As students explore the many career opportunities that exist in the global 21st century economy, they will also discover the requisite educational requirements for future careers. For schools and/or districts where Internet or computer access is limited, we provide some suggestions for hard copy career interest assessment methods.

Before we examine digital tools and websites, have students complete the Exercise on Career Possibilities (Exercise 8.1). A similar exercise will be presented later in this chapter after students have had the opportunity to expand their knowledge of possible career paths. Advisors/counselors should maintain copies of this exercise (digital or print) for future discussions.

Memo to Advisor/Counselor

The following presentation is in the voice of the advisor/counselor speaking to middle school students.

Today we are going to talk about what you want to do when you grow up. Let's think about possible careers you may want to pursue. Believe it or not, middle school is a perfect time to begin exploring career paths! How many of you have an idea of what your future career will be? Why are you interested in that career? Do you know what education is needed for that particular career?

After engaging in some initial discussion, the advisor/counselor should direct students to Exercise 8.1

Now, let's complete Exercise 8.1: Career Possibilities.

After students have completed Exercise 8.1, have a group discussion where students may share their responses. Remind students that this is an initial list of career paths,

EXERCISE 8.1 Career Possibilities

This exercise explores the time-honored question: "What do you want to be when you grow up?" Think about your answers. Try to name more than three career possibilities		
Possible Careers	Why this Career?	Education Needed

An Overview of Free Online Sites for Career Exploration (Print versions as noted) is provided for advisor/counselor consideration.

- EducationPlanner.org is a website that offers career interest surveys (also available in print format) for students as well as Career Exploration Activities. The Activities section provides browsing by either searching with keywords or by industry. This site also helps students find careers that match their skill sets .[7]

- Career Clusters Activity is an extension of Education Planner. This site allows students to explore career clusters: groups of jobs that share common themes or require similar skills. Additionally, this site predicts occupations that are likely to have more job openings in the future. Furthermore, this site has a section *It's Not All About Money* where students can explore occupations that they may genuinely enjoy. [8]
- Holland Code Career provides a free career quiz based on the Holland Code system helping students discover careers that fit their interests, talents, and aptitude. Students will get scores for six major job areas.[9] As Curry and Milson note, the Holland Code begins with an assessment of a person's interests and aptitudes, making it "ideal for middle school students because they love learning about themselves." The Holland Code presents a description of six Holland types: realistic, investigative, artistic, social, enterprising, and conventional with corresponding career possibilities.[10]
- Kuder Career Interests Assessment is available for students as young as sixth grade. Students respond to questions in an Interest Assessment that then leads them to their Top Career Pathways where they can explore career clusters, occupations, and majors, all based upon the results of their Interest Assessment.[11]
- *Myers-Briggs Type Indicator®* (MBTI®) is a career assessment tool based on an interest inventory assessment designed to help people as they explore career options. This test sorts individuals into 16 personality types based on four pairs of traits: Introversion/Extraversion, Sensing/Intuition, Thinking/Feeling, and Judging/Perceiving. The Myers-Briggs Type Indicator then matches these pairs to possible careers.[12]

- O*NET Interest Profiler provides career quiz questions to help students discover their interests and how they relate to careers. After completing career interest surveys, students then move to career exploration at MY NEXT MOVE[13] where they may search careers using keywords or by industry. This website also includes a search for apprenticeships through Apprenticeship Job Finder.[14]

  An added note here: Some middle school students may choose the apprenticeship route too early, particularly if they do not see themselves as a strong academically. Advisors/counselors should encourage students that they are on a journey that will help them become better students. Their self-concept is hopefully expanding in their awareness of their academic abilities. Also, advisors and counselors should explain that many apprentice programs require an associate's degree.
- Roadtrip Nation offers an online course that includes discovering career interests, exploring career possibilities, and conversations with people in given careers. This site uses a five point program:

  1. What you're interested in matters.
  2. Shed the noise and listen to yourself.
  3. Connect your interests to careers.
  4. Talk to a professional for advice.
  5. Overcoming roadblocks and taking action.

  RoadTrip Nation also offers multiple videos, stories, and podcasts for students. [15]
- The College Board's BigFuture provides a Career Quiz and responding careers related to the results of the quiz. Multiple careers are provided with descriptions of particular jobs, educational requirements, and income range. Videos where professionals in many fields discuss their

career work are also available. However, students must be 13 years old to register in the College Board's BigFuture, consequently limiting its use in middle school.[16]

- The ACT website allows students as young as 11 years old to create an ACT account. Once created, students may take career interest inventories that lead to the **Career Map** that matches their interests to exploration of possible careers and helps them to understand the day-to-day work that a particular career involves. This student account with the ACT will be used to access college exploration in a later chapter.[17]

8th grade students who have taken the **PreACT 8/9** receive Student Reports that delineate their Education and Career Journey based upon expressed career interests. This Student Report sends them to the **Career Map**. Similarly, eighth grade students in schools that administer the **ACT EPAS (the Educational Planning and Assessment System)** are guided to EXPLOREon the ACT website. However, it may be highly beneficial for these students to take the Interests, Abilities, and Values Inventories to further expand on their career interests.

Since the ACT provides valuable tools to students as young as 11 years old, career exploration and later college searches will focus on the ACT website. Ideally, the creation of an ACT account and the initial steps of career exploration should be done in Advisory or Group Counseling Sessions. If this is not possible, advisors/counselors should schedule ongoing group meetings to help students in this all-important process. The ACT website is an invaluable tool for middle school students as they come to understand themselves better, and at the same time, as they become increasingly engaged in the exploration of future career possibilities. This critical process takes time as students are involved in an ongoing career exploration that will hopefully broaden their self-concept.

Memo to Advisor/Counselor

Now, in the voice of the advisor/counselor helping students to establish an ACT account and to start initial career exploration…

What we are going to do today is create your account with the ACT. Many of you may know that the ACT is a college entrance exam, but what the ACT also offers is a way to explore careers based on your interests, values, and abilities. To do this, you need to establish an ACT account.

Getting Started…

Let's all go to the ACT website[18]

Here you will:

1. Create an account with ACT by using your email address and creating a password.
2. Upon signing in, go to the **Welcoming Page.** At the top of this page, select **Plans,** which leads to a page **exploring** your plans for the future.
3. You then select **Take Your Inventories.**
4. Once at the **Inventories Page**, you will take three Inventories: Interests, Abilities, and Values.
5. Each **Inventory** seeks to help you know more about yourself. These inventories are not about right or wrong answers. Simply answer the questions honestly! What you are doing is building a profile of yourself that will lead you to career areas on the **Career Map.**

    - The **Interest Inventory** consists of 72 questions and you select from three possible answers: Dislike, Indifferent, and Like.
    - The **Abilities Inventory** has questions about 18 abilities. Here you estimate where your abilities are in given areas by selecting:

        High–Upper 10%;
        Above Average–Upper 25%;
        Average–Middle 50%
        Below Average–Lower 25% and
        Low–Lower 10%

Advisors/counselors may find it helpful to caution students about overestimating or underestimating their abilities.

- The **Values Inventory** consists of 22 characteristics of occupations. You select from the following possible answers:

  Don't Want—I want to avoid work that has this;

  Not Important—I don't care if work has this or not;

  Somewhat Important—I hope to find work that has this;

  Very Important—I really want work that has this.

After you complete all three inventories, you are then directed to Explore Careers, which leads you to the Career Map.

As you look at the **Career Map,** you will see how the **Map** shows how Career Areas differ in their involvement with four basic work tasks:

People: People you help, serve, care for, or sell things to

Data: Facts, numbers, files, and business procedures

Things: Machines, tools, living things, and materials (like food, wood, or metal)

Ideas: Knowledge, insights, and new ways of expressing something (using words, music, etc.)

These four work tasks are the four "compass points" of the Career Map. For example, occupations in Career Area C (Management) mostly involve working with people and data, so it is located in the upper left part of the Map, as are other Career Areas involving people and data. On the other side of the Map, Career Area P (Natural Science & Technologies) mostly involves working with ideas and things.

Another key benefit of using the **ACT's Inventories** leading to the **Career Map** is that you may also view their results on the **Major Map,** a helpful aid as you begin college exploration in Chapter 9.

An Essential Next Step: An Important Page: My Summary

Click on the **PLANS** tab on the top of the **ACT** page. Then click on the **View Your Summary** tab. This will take you to your **My Summary** page. On this page, you will **see Career Areas that match two or more inventories** as well as **Career Areas that match one inventory.**

This **Summary Page** is an excellent way to explore the kind of work a particular career involves. For example, you may find that your inventories match multiple careers in Engineering & Technologies. When you click on that tab, it will open to multiple fields in engineering and technology. You will then click on a particular field such as mechanical engineer leading you to a page giving a broad overview of what a mechanical engineer may do. Here, you will see the following tabs: **Snapshot, Details, Training, Qualities, Salary & Outlook, and Related Careers**. Look carefully at each tab—this will give you a fuller picture of what it is like to be in this profession. On this page, you will discover educational requirements, salary, workplace, etc. Note that the **Training** tab talks about education needed as well as recommended high school classes.

Follow this procedure for all of the career paths that your **Inventories** show are of interest to you. This will take time, but you will learn about the work involved in careers that you may want to pursue. As you explore these career possibilities, make sure that you get a sense of the work involved. Try to imagine your future self on this career path!

Now that you have completed career interest inventories and have looked at some possible career paths, especially some careers that you may not yet have encountered, let's look at what you have discovered.

First, let's complete Exercise 8.2: Career Possibilities After Research.

EXERCISE 8.2 Career Possibilities After Research

Possible Careers	*Why this Career?*	*Is this Career New to You?*

Memo to Advisor/Counselor

After completing Exercise 8.2, students should compare Exercise 8.1 and Exercise 8.2 and note any differences in their answers on the two exercises. Advisors/counselors should maintain copies (digital or print) for future discussions.

Follow-up discussions with other students is an excellent way for students to share their career possibilities. Depending upon the size of the group, students may be divided into small discussion groups, or the advisor/counselor may conduct one large group discussion. What is essential is that students share their broadening career possibilities with each other and the group. This process should be repeated as students continue to use websites for career exploration.

After students have researched multiple career possibilities, they should complete Exercise 8.3: 5 Possible Careers.

In ongoing group discussions or individual meetings with students, advisors/counselors may then encourage students to use this information to motivate them to achieve greater academic success. What students have discovered about their interests may also impact their choices in classes and extracurricular activities as well as their pursuit of opportunities within their

EXERCISE 8.3 Five Possible Careers

List 5 Career Possibilities	*Why this career?*	*Education Required*
1.		
2.		
3.		
4.		
5.		

communities. Such early exposure to career options is certainly achievement motivating for this age group.

A final note here: *Students may find it helpful to retake the Inventories periodically, especially if they have new experiences in extracurricular or community involvement in career-oriented activities (environmental club, volunteering at an animal shelter or food pantry, etc.).*

Career Fairs

Career Fairs are frequently held in junior and senior years, but middle school students can clearly benefit from this activity. For schools where students have consistently engaged in Advisory or Group Counseling Sessions, a Career Fair is a wonderful opportunity to meet personally with professionals in multiple career paths.

For middle school students who have not had the opportunity to explore careers on a consistent basis, a Career Fair held once or twice a year may provide motivation to explore possible career paths. As part of the planning of the Career Fair, advisors/counselors should have pre- and post-meetings with their students. In the pre-fair session, students should complete either Exercise 8.1 or 8.2 depending on the timing/scheduling of the Career Fair and discuss what motivates their choices of possible career paths. In the post-fair session, students should complete Exercise 8.4 and discuss any new career possibilities revealed.

Career Fairs come in multiple formats from the full day model to the half day model to multiple virtual sessions. What is key here is that students interact with members of the community involved in multiple careers.

Some helpful planning sites include:

- https://www.schoolcounselor.org/Newsletters/January-2024/Engaging-Middle-School-Career-Development?st=IL[19]
- https://texasoncourse.org/media/2770/texasoncourse_careerfairtoolkit.pdf [20]
- https://sched.com/blog/career-day-ideas/ [21]
- https://www.teacherspayteachers.com/browse/free?search=career%20fair [22]

Looking at career possibilities can be an exciting and motivating experience for middle school students. As Michelle Shepherd comments in the *Illinois School Counselor*, "We host a career fair for our 4th- through 8th - grade students. Watching students get excited about jobs they have never thought about is rewarding and fun…….Career fairs are a lot of organization and planning, but they have a powerful impact on students."[23] Events such as Career Fairs have a way of bringing career exploration to life in the eyes of middle school students. As Michelle Shepherd comments, yes, there is much work involved, but the excitement in the middle school students is another affirming yes! This activity is a truly worthwhile endeavor that broadens the perspectives of middle school students.

After students have participated in a Career Fair, they should complete Exercise 8.4.

EXERCISE 8.4 Post Career Fair

List 3–5 Careers presented today that interest you	*What attracts you to this career?*	*What did you learn about this career today?*
1.		
2.		
3.		
4.		
5.		

Another activity that would show a middle school student what it is like to actually work in a specific career is to have a ***Shadow Day***. In this activity, students quite literally shadow or follow a person in a career field of interest to them for a day. While this activity takes planning to ensure that all students are matched with an appropriate person to shadow, this experience will likely broaden students' understanding of specific careers. Educators and parents alike should become involved in finding the right matches for students. If your school does conduct a *Shadow Day*, students should complete the following Exercise 8.5.

EXERCISE 8.5 What Did I Learn From My *Shadow Day* Experience

Student's Name

Name of the person you shadowed

Career area of the person you shadowed

Briefly, describe what you did over the course of the day.

What did you learn about this career field?

Are you still interested in this career field? If yes, explain why. If no, explain why.

One final exercise: As students continue their ongoing career exploration journey, why not have them interview a person in a career field that interests them? Family members, people in their hometown community, coaches, and neighbors, among others, may provide middle school students with interview possibilities.

Again, in the voice of the advisor/counselor...

Now, we are going to do one final exercise on careers. What you will do is interview someone you know who is in a career of interest to you. This person may be a family member, a neighbor, or a friend of the family. The main idea is that this person works in a career field that you want to explore more deeply. Remember that the career field does not have to be an exact match. For example, if you want to go into space exploration, it's going to be hard to find an astronaut, but interviewing an engineer may give you some insight into working in aerospace. Don't worry, if you can't find anyone to interview, we here at your middle school will be able to find a person for you to interview.

EXERCISE 8.6 Students Interview a Person in a Career of Interest to Them.

1. Name of Person
2. Occupation
3. Why did you choose this career?
4. What kind of education is necessary for this career?
5. How long have you worked in this career field?
6. Do you enjoy your work?
7. What do you like best about your work?
8. What do you like least about your work?
9. What advice do you have for students my age regarding careers?
10. Any comment that you might like to add?

We, as co-authors, have experienced the enthusiasm, the excitement, that career exploration ignites in our work with middle schoolers. This age group is open to future possibilities and to finding ways to experience careers through extracurricular activities such as volunteering at an animal shelter or a food pantry. As these middle schoolers become high school students, more class choices and activities will be available to them. In our advisory/group counseling system, the focus on career readiness may begin as early as second semester of sixth grade and continue into seventh/eighth grade. The natural flow of college exploration will have more meaning if students have initiated career searches first. As middle school students are now on a solid possible career path journey, we move to the next step: initial college exploration in Chapter 9.

Notes

1 Bandura, Albert. *Social Foundation of Thought and Action: A Social Cognitive Theory*. Prentice Hall, Inc, 1986.

2 Gaertner, Matthew and Katie McClarity. "Performance, Perseverance, and the Full College Readiness." *Educational Measurement Issues and Practice*, Summer 2015, Vol. 34, No. 2, pp 20–33.

3 Curry, Jennifer R. and Amy Milsom. *Career and College Readiness Counseling in P-12 Schools*, 2nd edition. New York: Springer Publishing Co., 2017, p. 240.
4 Curry, Jennifer R. and Amy Milsom. *Career and College Readiness Counseling in P-12 Schools*, 2nd edition. New York: Springer Publishing Co., 2017, p. 215.
5 Curry, Jennifer R. and Amy Milsom. *Career and College Readiness Counseling in P-12 Schools*, 2nd edition. New York: Springer Publishing Co., 2017, p. 219.
6 Field, Kelly. "The Path to a Career Could Start in Middle School." *The Hechinger Report*, November 2022. https://hechingerreport.org/the-path-to-a-career-could-start-in-middle-school/
7 "What's Your Learning Style?" *EducationPlanner.Org*. https://www.educationplanner.org/students/self-assessments/learning-styles
8 "Career Clusters Activity." *EducationPlanner.Org*. https://www.educationplanner.org/students/self-assessments/learning-styles
9 "Holland Code Career Quiz." *Truity*. https://www.truity.com/test/holland-code-career-test
10 Curry, Jennifer R. and Amy Milsom. *Career and College Readiness Counseling in P-12 Schools*, 2nd edition. New York: Springer Publishing Co., 2017, pp. 216–217.
11 "Students: How to Take Assessments in Navigator." *Kudor Help Center*. https://help.kuder.com/en/articles/6063433-students-how-to-take-assessments-in-navigator
12 "Strong Interest Inventory Tool." *The Myers-Briggs Company*. https://eu.themyersbriggs.com/en/tools/Strong-Interest-Inventory
13 My Next Move. https://www.mynextmove.org/
14 "Apprenticeship Finder." *ApprenticeshipUSA*. https://www.apprenticeship.gov/apprenticeship-job-finder
15 Roadtrip Nation. https://roadtripnation.com/experience/course/rtn/overview
16 "Big Future." *College Board*. https://bigfuture.collegeboard.org/
17 "ACT Student Career Interest Inventory." *ACT*. https://www.act.org/content/act/en/students-and-parents/high-school-success/what-to-do-after-high-school.html
18 "ACT Student Career Interest Inventory." *ACT*. https://www.act.org/content/act/en/students-and-parents/high-school-success/what-to-do-after-high-school.html
19 Shepherd, Michelle. "Engaging Middle School Career Development." *Illinois School Counselor, Illinois School Counselor Association*. https://www.schoolcounselor.org/Newsletters/January-2024/Engaging-Middle-School-Career-Development?st=IL
20 "Middle School Career Fair Toolkit." *Texas OnCourse*. https://texasoncourse.org/media/2770/texasoncourse_careerfairtoolkit.pdf
21 McTaw, Marvin. "40+ Career Day Ideas to Inspire Students." *Sched*, November 2024. https://sched.com/blog/career-day-ideas/
22 "Career Fair." *Teachers Paying Teachers*. https://www.teacherspayteachers.com/browse/free?search=career%20fair
23 Shepherd, Michelle. "Engaging Middle School Career Development." *Illinois School Counselor, Illinois School Counselor Association*. https://www.schoolcounselor.org/Newsletters/January-2024/Engaging-Middle-School-Career-Development?st=IL

9

College Exploration in Middle School

In Chapter 8, we focused on developing middle school students' awareness of the many career paths that they may find fulfilling. Now in Chapter 9, we seek to expand this awareness through an initial exploration of colleges. As students engaged in taking the inventories available on the ACT website that helped them to reflect upon their values, interests, and abilities, they came to understand themselves in a new way. For many middle school students, this heightened insight into themselves may have brought multiple possible career paths to their minds as they answer the age-old question "What do you want to be when you grow up?". As students engaged in career exploration, including understanding the nature of a particular career, its job description, and the education required, their perspectives have broadened, making them feel excited about their future selves. Now is a key time to build on that excitement and curiosity as students engage in a preliminary college search that further helps middle school students to think positively about their postsecondary options in this fast-paced, global economy.

A consistent goal of the advisory/group counseling systems has been to enhance the school environment, ensuring that all middle school students feel that they are part of a school

DOI: 10.4324/9781003597292-11

community where all involved—educators, parents, and students alike—are actively striving for middle students' growth. Initially, the emphasis is on establishing strong study habits and skills, the very foundation of the College and Career Readiness Standards. As students experience success in goal-driven tasks, particularly in academic achievement, the next aim of the advisory/group counseling systems is to make middle school students aware of postsecondary options through career exploration. As we enter this vital next step in broadening students' understanding of their future selves, the advisory/group counseling systems actively foster another factor: college-going beliefs. Research has found that helping middle school students broaden their perceptions of college is essential. A recent study "identifies eighth grade as a critical time in a student's education in which they make important post-secondary decisions. The research stressed that a student's academic achievement was not only influenced by intellectual abilities, but by their school climate. The purpose of this mixed methods study was to identify a correlation between school-based factors, specifically a student's sense of belonging, and college-going beliefs......... The results of this research will help educators improve programs, develop stronger relationships with students, and strengthen the curriculum to better prepare students for high school and beyond."[1] Of key importance in this research is the emphasis on school-based factors, particularly a student's sense of belonging and college-going beliefs. The advisory/group counseling programs are designed to promote both of these beliefs as students move through the middle school years. While not all middle school students will eventually attend a four-year college, what we are promoting is that education beyond high school is increasingly essential in the work environment. We want to foster the middle school students' sense of college-going beliefs so that they are open to multiple career paths and options as they enter the post-secondary stage of their young lives. Now is a vitally important time to introduce a preliminary college search as middle school students hopefully become more engaged in college-going beliefs.

Post-secondary options are interconnected as we introduce college searches based upon career exploration. According to the

Hechinger Report, preparing students for a changing workforce is an important goal in career exploration, a goal that flows quite naturally into an early college exploration. As noted in Chapter 8, "A growing number of states and school districts now require students to take career exploration classes in middle school. Others offer introductory courses in specific careers, like engineering or robotics.... Dallas Independent School District...... has long offered career exploration courses to its seventh and eighth graders. But this year it expanded one of the classes, based on a curriculum from the nonprofit Education Opens Doors, to every middle school in the district. Brian Lusk, the district's chief of strategic initiatives, said school leaders wanted to ensure that all students were prepared to make informed decisions about their paths in high school and beyond. "Equity is important to us, he said."" Furthermore, "Students are less stressed out in the middle grades," said Stephanie Simpson, CEO of the Association for Middle Level Education, a nonprofit that supports middle school educators. "They can explore and take some risks, with fewer immediate consequences."[2] As students consider possible careers, early college research is an integral and natural ongoing part of this exploration.

Another argument for early career and college exploration in middle school is that these students, as they move on to high school, will be more open to taking classes and becoming involved in extracurricular activities that reflect their future aspirations and goals. Nancy Deutsch, a University of Virginia professor who is the director of Youth-Nex: The UVA Center to Promote Effective Youth Development, is leading an effort to remake middle schools. She believes that the early teen years may be the ideal time to start this research, as early adolescence is a pivotal time for identity development when young people are seeking to define themselves and quite literally trying on their "future selves." Furthermore, Deutsch believes that "the early teen years are also a stage when students are especially vulnerable to "identity foreclosure," or the walling off of certain options, such as a STEM career, as not for them... By catching students before they foreclose, schools may be able to convince more female students to consider computer science, for example."[3] Once again, this argument affirms that middle school is an ideal time for students to consider both career and college future possibilities; indeed, it is an opportune time for this all-important early career and college exploration.

Moreover, the changing picture of American secondary education also relates to the importance of middle school career and college exploration. Particularly in large metropolitan areas, middle school students may have the option of applying to specialized high schools where the emphasis may be on science, engineering, or American Studies, among others. For example, the New York Public School System offers eight specialized high schools where admission is based upon testing. The ninth specialized high school for performing arts is based upon audition and school record. In addition, multiple high school options for secondary education exist in NYC that are not related to admission testing. Furthermore, in 2021, NYC discontinued zoned schooling for high school, allowing students more choices in high school placement.[4] In Dallas, eighth graders must choose one of five endorsements to focus on in high school—among them, STEM (science, technology, engineering, and math), business and industry, and the arts and humanities.[5] Currently, Indiana now requires all eighth graders to take a series of self-assessments through the state's online career explorer or a similar web tool. The results are shared with guidance counselors, who help students match their interests, strengths, and values with one of three paths: employment, enrollment, or enlistment. Similarly, Delaware is in the process of writing standards for career and technical education in the middle grades, after finding that middle schoolers are often making uninformed decisions about which high school to attend. Indeed, Virginia has students begin work on an "academic and career plan portfolio," which includes information about their interests, values, and skills, as early as elementary school.[6] These programs reflect the ongoing, developing awareness of the need for career and college exploration in middle school. (See Chapter 10 for more information about high school options.)

How can middle school students begin to make intelligent choices without career and college exploration? As the Hechinger Report also states, practical reasons exist to start career and college exploration in middle school. "High school is far too late to begin this conversation with young people," said Kyle Hartung, an associate vice president with Jobs for the Future, a nonprofit that offers a career exploration curriculum for schools and after-school programs. In a pair of recent surveys by American Student Assistance, a nonprofit focused on career readiness, roughly two-thirds of high school graduates said they would have benefited from more career exploration in middle or high school, and 80% of high school guidance counselors said their students were "overwhelmed" by decisions about college and career.[7] Once again, what is essential to this discussion is that career exploration flows naturally into an initial research of colleges.

Simultaneously, as middle school students explore colleges in this chapter, their sense of belonging in postsecondary education will expand, motivating them to strive for greater academic success. Clearly, middle schoolers are at a prime time in their development where they are open to exploring the possibilities of college study and future career paths in this rapidly changing, global workplace. Life in middle school will pass quickly, and eighth graders will not only be moving on to high school, but also perhaps competing for placement into competitive secondary environments. Now is the opportunity for us as educators to foster their sense of belonging in the postsecondary world of college study.

A side note: As coauthor (Kathleen Costantini), I have personally experienced how the middle school years are an ideal time to explore careers and colleges. During the COVID pandemic, I had the opportunity to work with a young middle school student, Meagan. Despite the isolation of the pandemic, Meagan always managed to be upbeat in our Zoom conferences. What we were discussing was not schoolwork; Meagan was and continues to be an actively engaged student, clearly possessing solid study habits and skills. We used this time for career exploration and an early college search. As Meagan took the ACT's Inventories, we both clearly saw how her interests, abilities, and values inventories pointed her in the direction of the health sciences. Fast forward to high school, Meagan opted for challenging Advanced

Placement classes, and during her summers, worked as a counselor-in-training and later a counselor in her town's summer camp for young children. Her empathy made her a natural counselor as she helped young children have a positive, enjoyable camp experience. During the summer between her junior and senior years, Meagan engaged in a precollege program at a university where she was formally introduced to many facets of the health studies. Meagan, now a senior, is awaiting admissions decisions and clarifying in her own mind which college or university will best enhance her growth in the health sciences field.

Memo to the advisor/counselor...

Just as students examined possible careers in Chapter 8, our aim in this chapter is to promote an initial college search related to students' career interests. Inherent in this preliminary college exploration is our goal to promote awareness of higher education in middle school students. The emphasis in this chapter will be on the **ACT's college search**, which allows students as young as 11 years old to establish accounts to explore higher education. (To access the College Board's college search, students must be 13 years old and may be used by older middle school students.) We want middle school students to discover the educational pathways to pursue their career possibilities. Also included in this chapter is a preliminary overview of financial aid and scholarship opportunities. Our purpose here is to show all students that postsecondary education is a viable option for them despite family income status. Advisors/counselors will be able to assist in this exploration by using the charts and exercises provided in this chapter.

Again, the emphasis is on an initial search of colleges and universities to heighten student awareness of the possibilities ahead!

This section is presented in the voice of the advisor/counselor speaking to students.

Now that you have researched possible careers, it is time to move on to the next step: college exploration. Remember, this is initial research of some colleges that may be of interest to you. Just as you looked at possible career paths, now we are going to examine how a college education, at both community colleges and four-year colleges, may help you pursue a career of your choice.

Before we begin exploring majors and colleges, let's review Exercise 8.3 about your possible career paths. (If a student has not completed 8.3, now is a perfect time to complete this essential exercise.)

EXERCISE 8.3 Possible Careers

List 5 Career Possibilities	*Why this Career?*	*Education Required*
1.		
2.		
3.		
4.		
5.		

Look at your choices of possible careers and why you are interested in these career paths. Now examine the necessary educational requirements for your career possibilities. In our global economy, most careers minimally demand some form of higher education. Some careers, such as doctors, lawyers, teachers, counselors, etc., require advanced educational degrees. Keep in mind that many apprenticeship programs require associate degrees.

Now, let's move on to our initial college search!

Using the **ACT,** let's begin an exploration of colleges. For those of you who do not yet have an ACT account, go to: https://my.act.org/account/signin?location=https://my.act.org[8] to create an account. Simply click on the **Create MyACT** account bar, then enter an email address (user) and a password. If you have already used the career exploration of the **ACT,** you will now be able to investigate possible majors and colleges available on the **ACT's website.**

As we explore colleges related to your career interests, we must first look at what majors are required for various occupations. First, what is a **major?** In college, you will select a major area of study to prepare you for future career possibilities. That concentrated area of study is your **major,** and it is key to exploring colleges. Don't worry if you do not feel certain about a particular area of study now. Many college students do not initially declare their majors until they have taken some of the college's required classes.

Remember, a world of opportunity is ahead of you! What we are doing now in middle school is helping you become more aware of the possibilities ahead for a fulfilling career and college experience.

So now, let's begin:

- You have already taken the **ACT's Inventories** that led you to the **Career Map** that helped you to complete Exercise 8.3 on Five Possible Careers. If any students have not yet completed the **Inventories,** now is the time for them to finish this all-important step.
- Sign in to **your MyACT.** On the **Welcoming Page,** click **Plans** in the upper bar.
- On the **Start Here** page, click on the **Find Majors** tab.
- Next, click on the **View Major Map** in the right corner of the page. The **Major Map** is designed to help you find

the majors in college that match your interests. This map organizes majors into 13 groups.

- Take a tour of **Major Map.**
- Once on this page, you will see a **Major Map Key**: **Major Group** is in black while **Major Groups** that match your **Interest Inventory** are in blue.
- Click to explore the **Major Group.**
- Most importantly, explore the **Major Groups in blue** that **match** your **Interest Inventory.** Here you will find the majors that suit your interests.

As you select a group, you will see the majors available in that area of study. Next, select a major to learn more about topics of study, degrees, related occupations, and more. Looking at all of this information will take time! But keep in mind, what you are doing is looking at possible futures for yourself!

A note to the advisor/counselor ... Exploring the Major Map will take time, but it is an essential step in expanding how middle school students view future career pathways.

Now that you have had some time to explore possible majors that will lead to career paths, let's complete **Exercise (9.1) Matching Possible Majors with Possible Careers.**

EXERCISE 9.1 Matching Possible Majors with Possible Careers

3 Possible Majors	*3 Possible Careers*

Remember that you can always visit the **View Your Summary bar under Plans** to review your **Careers, Majors, and Schools** as you continue your explorations!

Memo to the advisor/counselor:

Students exploring majors leading to career paths is an ideal time for individual meetings with the advisor/counselor and/or small group discussions where students can articulate the all-important **WHY** behind these majors and career preferences.

Again, in the voice of the advisor/counselor …

Now that you have researched some possible majors that you may want to pursue, let's first have an overview of higher education. When you are in high school, you will delve into the different kinds and levels of higher education in much greater detail. For now, let's look at the basics of higher education…. the where and what you study after you graduate from high school.

Note to the advisor/counselor...

The following discussion of higher education may be modified according to the age and readiness of the middle school student group.

What we are exploring now are **four-year colleges and universities** where you can obtain a **college degree.** Keep in mind that many students attend community colleges to earn an **associate degree (designed to be completed in two years)** and then transfer to a four-year college or university to complete their study to obtain a **college degree**. Also, some **apprenticeship programs** are conducted through community colleges resulting in an **associate degree.** Even if you think that a four-year college is not in your future, explore colleges now. We live in a rapidly changing, global economy. Keep your options open!

A note to the advisor/counselor: Although students are not yet ready to actually pursue college options, they should, even at this point in middle school, be aware of the statistics regarding the length of time that it currently takes to complete an associate degree. "According to the American Association of Community Colleges, 66 percent of students attend part-time. Studying part-time, it may take you (a student) anywhere from **48 to 72 months** to earn an associate degree."[9] Students in these programs are usually working part-time as they study. In essence, a majority of students take four–seven years to complete this degree. These statistics may not apply to apprentice programs where employers are involved with directing the students to complete a given apprenticeship.

Some important definitions:

Liberal Arts Colleges are colleges that offer a broad overview of the arts, sciences, and humanities.

Here are **some of the majors** that you may choose to study:

- Art history, fine arts, acting, classics, philosophy.
- Computer science, geological sciences, marine science.
- Economics, geography, linguistics, sociology, psychology.
- Education, early childhood, elementary, and secondary.
- English, writing component, foreign language.
- Language, literature, and writing.
- Math and sciences such as astronomy, biology, chemistry.
- Nutrition, physical science, physics.
- Social sciences, government, history, anthropology.

Liberal Arts Colleges tend to be smaller institutions. While they have some lecture classes, more emphasis is on classroom teaching and direct interaction between professors and students. Their smaller campus environment promotes social and cultural opportunities. However, not all liberal arts colleges are small; always check the school population to see how small or large the college is.

Universities are institutions of higher education that are composed of more than one college where students may earn a college degree.

For example, a typical university will consist of the following colleges:

- Liberal Arts.
- Engineering/Architecture.
- Education.
- Business.
- Science.

Universities are usually medium to large-sized institutions, many with name recognition. Frequently, professors teach lecture classes with smaller discussion groups led by teaching assistants. The university campus may seem like a city in itself, and similar to liberal arts colleges, provides many diverse social and cultural opportunities.

Another fact to keep in mind: Names can be a bit misleading: Boston College, Dartmouth College, and the College of William and Mary are all universities but have chosen to keep their original names for historical purposes.

So now that you have a better understanding of higher education, let's return to the **ACT website** to explore colleges related to your major and career interests.

- Sign into your **MyACT** account, and once again, go to the **Plans** tab. Once on the **Plans** page, go to **Explore Schools** to find colleges and universities related to your interests and possible majors.

On the **School Search** page, you will see a top bar that allows you to search schools by name, major or state. Below on the right column, you may search by:

- Major or Degree
- Location or School.
- Admission Difficulty.
- Annual Cost.
- Student Body.

For now, let's focus on the **Major or Degree, Location and School.** To begin, **choose three colleges that are in your state or three colleges that you know or recognize.** Remember, this is a basic introduction to exploring colleges. As you enter high school, you will have a broadening view of the choices ahead of you!

Now, using the **Search by school name, major or state bar at the top of the page,** explore three colleges of interest to you. Look at the size of the student body, location, and majors offered.

Memo to advisor/counselor....

Remind students that this is a very initial search of colleges to give them an idea of the possibilities ahead of them. Again, just as in the **Major Search**, this preliminary exploration of colleges will take time as students explore college possibilities.

Again, in the voice of the advisor/counselor....

Now that you have had some time to research colleges, let's complete

Exercise (9.2) Matching Possible Majors with Possible Colleges.

EXERCISE 9.2 Matching Possible Majors with Possible Colleges

List 3 Possible Majors	*List 3 Possible Colleges*	*Size and Location of College*	*What did you like about this college?*
1.			
2.			
3.			

Students exploring possible majors and colleges is an opportune time for individual advisor/counselor meetings with students and/or small group discussions where students may share their initial college search results.

Now, we have one final exercise to do. Remember that you interviewed a person in a particular career of interest to you (see Chapter 8.). Why not speak to someone who has attended college, majoring in an area of interest to you? If you can't find a person with a similar major, you may still do this exercise even though the **major** of the person being interviewed may be different from what you want to study. Just as in the career exercise, this person may be a friend of the family, a family member, a teacher at your school, etc.

After completing this exercise, have students share their interview results in a group discussion. Hopefully, students will have positive remarks to share. Even if students interview a person

EXERCISE 9.3 Students Interview a Person who Majored in an Area of Interest to Them.

1. Name of Person
2. Occupation
3. What was your major in college?
4. What career were you pursuing?
5. Did you attend community college? A four year college?
6. How long did it take you to achieve a degree?
7. Did you enjoy your college experience?
8. Did your degree help you achieve a career that you wanted?
9. What advice do you have for students my age regarding college?
10. Any comment that you might like to add?

who has had a negative collegiate experience, understanding the why behind this reaction is important. Finally, remind students that this is a very early college search, one that will continue in high school.

A final note: Chapter 10 talks about helping students find a good high school match for them. Particularly in large urban areas, eighth graders may apply to high schools that specialize in subject areas of interest to them—all the more reason for middle school students to be involved in career and college exploration!

Again, in the voice of the advisor/counselor….

Now that you have explored possible careers, majors, and colleges, let's have a broad overview of essential financial aid and scholarship programs. You are still in middle school: what we want you to know is that no matter your family income, college is an attainable goal for you!

Now some terms to know:

1. When you are in high school, your parents may file a **FAFSA**, or Free Application for Federal Student Aid, that measures your family's ability to pay for college.
2. The **FAFSA** then establishes your **Student Aid Index** toward college expenses.
3. Federal student aid may be in the form of grants, work-study plans, or loans. (Grants are gifts, work-study usually involves on-campus employment, and loans must be repaid.)

4. Nearly all states have some form of need-based financial aid. The states base their programs on the **FAFSA's Student Aid Index.**
5. Colleges and Universities also award need-based financial aid to students.
6. Merit-based scholarships are another form of financial assistance for students who qualify. These awards are not tied to financial need but are a recognition of academic achievement as well as special talent areas such as dance, music, art, and sports.

Also keep in mind that students who take the **PSAT/NMSQT** in junior year may qualify for financial awards for college through the **National Merit Scholarship Program.** Awards are given to finalists in all 50 states. Another part of this program is the **National Hispanic Recognition Program,** which honors the highest-scoring Hispanic students taking the **PSAT/NMSQT.**

Also, most state colleges and universities will have lower tuition costs than private schools. These lower rates apply to residents of that state. For example, if you live in the State of Michigan, the University of Michigan has different tuition rates for residents and nonresidents.

Remember, our purpose in exploring majors and colleges is to make you aware of the possibilities ahead. The brief highlights of financial aid and scholarships are to assure you that you will be able to pursue your career path.

A Final Note: Keep an open mind! You have many opportunities through classes, extracurricular activities, community involvement, etc. to broaden your perspective about your career and college searches. This is an initial step on your journey of college exploration!

Notes

1 Steiner, Jamie L. *The Relationship between Middle School Students' College-Going Beliefs and Sense of Belonging in School (2011).* Education Doctoral, Paper 39, Fisher Digital Publications, St. John Fisher University. https://fisherpub.sjf.edu/education_etd/39/

2 Field, Kelly. "The Path to a Career could Start in Middle School." *The Hechinger Report,* November 2022. https://hechingerreport.org/the-path-to-a-career-could-start-in-middle-school/
3 Field, Kelly. "The Path to a Career could Start in Middle School." *The Hechinger Report,* November 2022. https://hechingerreport .org /the -path -to-a -career -could -start -in -middle -school/
4 NYC Public Schools. https://www.schools.nyc.gov/
5 Dallas Independent School District. https://www.dallasisd.org/
6 Field, Kelly. "The Path to a Career could Start in Middle School." *The Hechinger Report,* November 2022. https://hechingerreport.org/the-path-to-a-career-could-start-in-middle-school/
7 Field, Kelly. "The Path to a Career could Start in Middle School." *The Hechinger Report,* November 2022. https://hechingerreport.org/the-path-to-a-career-could-start-in-middle-school/
8 "My ACT." *ACT.* https://my.act.org/account/signin?location=https://my.act.org
9 "How Long Does It Take To Get an Associate Degree?" *Coursera,* August 2025. https://www.coursera.org/articles/how-long-does-it-take-to-get-an-associate-degree

10

Articulation to High School

Finding Strong Matches in an Expanding World of High School Possibilities

Tricia is an eighth grader who is taking Algebra in her middle school at 7:30 a.m. before the regular school day starts. Her teacher, who works for an insurance company but comes in especially to teach this class, has suggested she try out for a high school that specializes in Mathematics and Technology. Every fall in the United States, eighth graders and, in some cases, sixth or seventh graders begin to explore and apply for high school placements, whether for specialized schools or regular public, private, or parochial schools. With the explosion of charter schools, specialized schools, and the closing of poorly performing schools nationwide, getting into high school, knowing the choices, and most importantly, being prepared to try for a particular high school has become a crucial part of the work of middle school counselors, teachers, and administrators.

Helping middle school students and their families navigate the world of high school placement may be daunting indeed. For many students in the United States, their choice will be the public high school in their local area; in other words, their zoned public school. For others, exploring private and parochial schools may be an option. In larger school districts, specialized high schools may be a possibility. Essential to this search for the best matches for students to achieve success in high school is the academic

DOI: 10.4324/9781003597292-12

advisement needed to enable middle school students to have choices of high schools where they can matriculate and be successful—a truly demanding endeavor.

Nationwide, school districts vary dramatically from the large urban centers to suburban areas to more rural sections of the country. Knowing the high school options in given districts is key to helping students find their best matches. Particularly large urban areas provide students with multiple options for secondary education. Here, we briefly look at three large urban school districts.

The **Chicago Public Schools** offer eighth graders multiple options for high school.

Types of High School Choices

- **Neighborhood Schools:** Every student has a designated local school they are guaranteed to attend.
- **Selective Enrollment:** For academically advanced students, requiring testing for rigorous college prep programs (e.g., Walter Payton, Northside Prep).
- **Magnet Schools:** Specialized schools focusing on specific themes (e.g., STEM, Fine Arts).
- **Career Academies:** Offer hands-on learning and career exploration in fields like Business, Agriculture, and Finance.
- **Charter Schools:** Independent public schools with unique models, also part of the choice system.

How Students Choose

- **Go CPS Application:** The centralized platform to explore and apply to up to 20 programs across different schools. [1]
- **Ranking:** Students rank their chosen programs, and offers are made based on program-specific selection processes (lottery or points).

- **Beyond Neighborhoods:** A large majority of eighth graders choose a high school outside their attendance zone.

Similarly, the **Dallas Independent School District** (Texas) offers students over 100 choices for high school. Dallas ISD aims to provide these options to all families to keep them within the district, expanding opportunities beyond traditional neighborhood schools.

Types of Choices Available

- **Collegiate Academies**

Collegiate academies are designed to serve incoming ninth-grade students who have the determination to attend college, are first-generation college students, and have been historically underrepresented in higher education.

- **Magnet Schools**

The district magnet school system is made of magnets for high school students, academies for middle school students, and vanguards for elementary school students. The schools provide career-oriented or specialized curriculum in addition to academic subjects.

- **Two-Way Dual Language**

Dallas ISD is creating global citizens everyday by providing our students with the opportunity to learn two languages. English speakers can learn Spanish and Spanish speakers can learn English within a program called the "Two-Way Dual Language Program."

- **Career Institutes**

Expand career options for highly motivated students who want to enter the skilled trades and earn a living wage right out of high school. The institutes offer courses leading to certificates in areas such as aviation, plumbing, electrical, cybersecurity, construction and much more.

How It Works

- **Application-Based:** Students must apply to these choice schools and programs.
- **Merit and Lottery:** Admission is often based on merit (like academic performance or auditions) or a lottery system, especially for popular programs.[2]

Students in the **Los Angeles Unified School District (LAUSD)** have significant choices beyond their neighborhood school.

Types of Choices Available:

- **Magnet Schools and Programs:** Specialized programs focusing on themes like Arts, STEM, Liberal Arts, or Media, accessible through an application process. **Zones of Choice (ZOC):** In participating areas, students can choose from several high schools within their zone, offering more options than traditional neighborhood assignments.
- **Charter Schools:** LAUSD also offers charter school options, providing alternatives to traditional public school.

How Students Apply:

- **LAUSD Choices Portal:** Students apply to various programs and schools during specific application windows, usually starting in October.
- **Multiple Applications:** Students can apply to several Choices programs and schools.

- **Accepting Offers:** Students can only accept one Choices placement, and accepting one automatically removes them from others.[3]

Nationwide, school districts are responding to the needs of students for placement in a high school where they may pursue areas of most interest to them. As counselors, teachers and administrators nationwide work to assist students and their families to find the best academic possibilities for high school in their districts, it is essential to keep in mind that what is most important is finding the best matches for students. We want students to find a secondary school where they will thrive and be challenged. Increasingly these options will reflect a students' early sense of career awareness as well as postsecondary options, as we have previously explored in the advisory/group counseling sessions.

How to Present High School Options in the Advisory/Group Counseling Program

For purposes of the advisory/group counseling sessions, let's now examine the public school system in New York City in greater detail. In the largest public school system in the nation, New York City offers many high school opportunities. Much of the statistical information is for the advisor/counselor. What students need to know are the appropriate options and possibilities for them as individual students. In other words, what high school in this very large district will enhance their growth? At which schools are they a viable candidate? Looking at how the New York City Public School System offers various high school possibilities may help counselors, teachers and administrators examine their own school districts' options as they advise middle school students in this all-important next step of their education. Exercises may easily be adapted to individual school districts. Once again, not all of the statistical data is necessary for use in the advisory/group counseling programs. Hopefully, this information may assist advisors and counselors in other school districts, particularly urban areas, as they assist students in finding the best high school option for them.

A Closer Look at the New York City Public School System for Advisors/Counselors

In 2024–25, there were 906,248 students in the NYC school system, the largest school district in the United States. Of those students:

- 16.8 percent of students were English Language Learners.
- 22.4 percent were students with disabilities.
- 75.3 percent were economically disadvantaged.

Race or ethnicity: 42.3% Hispanic, 19.3% Black, 18.7% Asian and Pacific Islander, 16.1% White, 2.0% Multi-Racial, and 1.2% Native American.

There were 1,597 schools within NYCPS as of the 2024–25 school year and 281 charter schools.[4]

Another view of the complexity of the NYC public system's population:

- There are more ELL (English Language Learners) students in NYC (161,896) than in Chicago and Houston combined. The size of NYC's ELL students alone is greater than the size of 90% of school districts in the country.
- NYC has almost 100,000 more students with disabilities than LA, Chicago, and Houston combined. 20% of NYC's students are characterized as having some type of disability, compared with 12% in LA, 14% in Chicago, and 7% in Houston.

The large numbers of students—and their diversity—create inherent challenges for implementing education programs and policies in NYC. For instance, school and district officials have to communicate with families who speak over 182 different languages.[5]

What are the High School Possibilities for Middle Schoolers?

The NYC Public School System has 533 high schools. How to begin to navigate the NYC public school system? How do you find the right school?

1. Zoned schools are the high schools associated with a student's address. In 2021, the New York City Department of Education (DOE) eliminated geographic restrictions for admission to high schools. This means that students can apply to any high school, regardless of their borough, district, or neighborhood. Students can apply to high schools through the New York City Public Schools website.[6]
2. Charter schools: Several sites exist to apply to Charter Schools as individual schools may have students apply through their sites related to specific schools. See the NYC Department of Education for further sites. Interested families may also contact specific Charter Schools for more detailed information.[7] NYC public charter schools continue to see growth in enrollment as families seek high-quality public school options. Since New York City's first charter school opened in 1999, the charter school sector has grown to encompass 281 schools across all five boroughs, serving an estimated 148,089 students.
3. Specialized High Schools: NYC has nine Specialized High Schools The nine Specialized High Schools are one way that New York City supports the educational needs of students who excel academically and/or artistically, including English language learners and students with disabilities. An average of 25,000 students take the Specialized High Schools Admissions Test (SHSAT) each year to apply to specialized high schools in New York City.[8]

A special note to advisors/counselors: As coauthors and educators working in New York City, we have both experienced the effects of counselors and teachers advising students against applying to competitive schools. A case in point is young Tommy who was advised against taking the Specialized High School Test because his teachers did not believe that he would be successful. When Tommy and his parents came to this coauthor (Kathleen Costantini) as a private client, I agreed with the family that Tommy was a solid candidate. Just as high school students study for the ACT or the SAT, Tommy worked hard to prepare for the Specialized High School Exam. I still remember the phone call that I received from an excited Tommy. Not only was he accepted to one of the specialized high schools, but his acceptance was to his first choice school. Indeed, Tommy became the first student from his grade school to be accepted at the highly ranked Stuyvesant High School. While his teachers and counselors may have been trying to prevent Tommy from experiencing rejection at a young age, their advice may have prevented him from partaking in a wonderful educational opportunity. Tommy thrived at Stuyvesant and went on to further his studies in college.

What are the Qualifications for the Specialized High Schools?

Necessary qualifications: To register for the SHSAT for admission to the Eight Specialized High Schools and to audition for LaGuardia, a student must be both of the following:

- A NYC resident.
- A current eighth-grade student or first-time (not repeating the grade) ninth-grade student.

Eight Specialized High Schools—Take the SHSAT to Apply

Eight Specialized High Schools use the Specialized High Schools Admissions Test (SHSAT) to determine who can attend these schools. During the registration period for the SHSAT, students indicate their preference order for the testing Specialized High Schools. Students are then considered for an offer to these schools by taking the SHSAT.

CHART 10.1 EIGHT SPECIALIZED HIGH SCHOOLS THAT REQUIRE THE SHSAT

Eight Specialized High Schools Requiring the SHSAT

- Brooklyn Latin School.
- Brooklyn Technical High School.
- Bronx High School of Science.
- High School for Math, Science and Engineering at City College.
- High School of American Studies at Lehman College.
- Queens High School for the Sciences at York College.
- Staten Island Technical High School.
- Stuyvesant High School.

The SHSAT is not used to determine admissions to the ninth specialized school, Fiorello H. LaGuardia High School of Music & Art and Performing Arts. This school uses an audition process for admission. At La Guardia, students complete two parallel paths of study: conservatory-level pre-professional training in the arts and college-preparatory academics. Students can apply to LaGuardia by submitting a portfolio and auditioning for up to six different studios:

- Dance.
- Drama.
- Fine & Visual Art.
- Instrumental Music.
- Technical Theater.
- Vocal Music.

LaGuardia High School scores auditions for each studio, and offers are based on audition score. Students receive an offer to each studio for which they have a qualifying audition score, meaning students can receive up to six offers to LaGuardia. To receive an offer to LaGuardia, students must have a documented final grade from the last school year of at least 65 or the equivalent in each core subject (ELA, math, science, social studies).

The SHSAT is the only criteria used to admit students to the eight testing Specialized High Schools; no other criteria are used.[9] As high school students prepare for the ACT and the SAT exams, students must also prepare for the SHSAT. Preparation may begin during the summer before eighth grade and continue right up to the test.

CHART 10.2 HOW TO PREPARE FOR THE SHSAT

1. Understand the test format. Visit https://www.schools.nyc.gov/learning/testing/specialized-high-school-admissions-test This site provides two Sample Tests plus answers and explanations.
2. Become familiar with the sections of the SHSAT: Length of the test is 180 minutes.

    Part I—English Language Arts—57 questions that include Revising/Editing (1–9) and Reading Comprehension (10–57).

    Part II Mathematics—57 questions that include Grid-In Questions (58–62) where students write in their answers and Multiple Choice Questions (63–114).

Multiple book versions and online prep are available.

Some middle schools and local groups conduct prep sessions. Investigate what help is available.

Qualified students, such as those with an Individualized Education Program (IEP) or English Language Learners (ELLs), can receive extended time on the SHSAT.[10]

Offers to the eight testing Specialized High Schools are based on each student's SHSAT score, school preference order, and the number of seats at each school. These offers are made to each school in descending SHSAT score order until all seats at each school are filled.

> Here's how that works:
>
> The student with the highest SHSAT score receives an offer to their first choice school. The student with the next highest SHSAT score receives an offer to their first choice school. This continues student by student in downward score order until a student's first choice school does not have any remaining seats. That student is then considered for their second choice school. If there are remaining seats at the second choice school, that student receives an offer there. If the second choice school has no remaining seats, that student is considered for their third choice school, and so on. This continues until all seats at all Specialized High Schools have been filled. If a student doesn't get an offer to their first choice school, it is because all seats at that school have been filled by students with higher SHSAT scores. Similarly, if a student gets an offer to a non-first choice school, it is because all seats at schools they listed higher have been filled by students with higher SHSAT scores (e.g., if a student gets an offer to their fourth choice school, then seats at their first to third choice schools are filled with higher scoring students.)
>
> If a student doesn't get any offer to a testing Specialized High School, it is because all seats at all the schools they listed on their SHSAT registration are filled by students with higher SHSAT scores. Students are not considered for offers to schools that are not listed on their SHSAT registration.[11]

New York City Public Schools also have specialized high schools where admission is NOT connected to the SHSAT. Here are just a few examples. Explore the options available!

- Manhattan Hunter Science High School uses a screened admissions process and requires an essay, which accounts for 30% of the admission decision. It is open to students in all five boroughs. MHSS is an early college high school that partners with Hunter College. The school has a 98%

graduation rate, with 92% of students going on to four-year colleges. In addition to its strong Math and Science curriculum, the school offers extensive AP and college-level courses across disciplines. MHSS has an excellent student-to-teacher ratio of 18:1 and places an emphasis on support for its students' success.[12]

- Academy of American Studies in Long Island City offers multiple advanced placement courses. Admission is based on NYC DOE admissions policy and procedures.[13]
- The Frank Sinatra School of the Arts is an arts high school in Astoria, Queens that combines strong academic performance including advanced placement classes with emphasis on the performing arts. Founded by Tony Bennett and his wife Susan Bennett, a NYC teacher, the school is located in Tony Bennett's home neighborhood and named for his best friend, Frank Sinatra.[14]
- New Explorations in Science, Technology & Math School is located in Manhattan. The school uses a screened admissions process and requires an essay which accounts for 50% of the admission decision.[15]

In addition to public school options, NYC is also home to many private schools. Students may want to explore Catholic high schools and private independent schools. These schools require entrance exams. While these schools are tuition-based, many offer scholarships and financial aid.

Now as counselors, teachers and administrators examine the high school options available in their school district, let us look at how to present this information to students. This discussion may begin in seventh grade, but no later than early in the first semester of eighth grade.

Again, in the voice of the advisor/counselor....

Now, it is time for us to discuss what you would like in your future high school. Are you seeking a specialized high school that requires an entrance exam? A charter school? Your zoned or neighborhood school? Think about the size of the school, the location—how far will you have to travel?

Let's begin an initial exploration of high schools in our school district. School options and descriptions are available on the district's website.

Exploring schools may take time, but it is an essential step in finding the right high school match. As students research possibilities, have frequent group discussions so that all students learn about as many secondary options as possible.

After students have explored several high school options, have them complete Exercise 10.3.

EXERCISE 10.1 Initial High School Possibilities

List 5 High School Options	*What Attracts You to this School?*	*What is the Admission Process?*

After students have completed this exercise, break students into small groups or have one large group discussion where students discuss their high school options. Advisors/counselors should maintain hard or digital copies of Exercise 10.3.

Again, in the voice of the advisor/counselor...

Some schools have an Open House for prospective students to visit and learn more about the school. Make sure to check to see if there are any Open Houses of interest to you! Another important factor, especially if you are applying to a high school that requires an admission test, how do you prepare for the test? (Here, counselors may be able to suggest methods of preparation—a published test book for the exam, tutoring opportunities, etc.).

After students have explored more high school possibilities, have them complete the following exercise.

EXERCISE 10.2 Five High Schools of Solid Interest to Me.

List 5 High Schools I like	*What kind of school is this? Specialized, charter, neighborhood, etc.?*	*Admission Criteria—If testing or portfolio is required, how do I prepare?*

Again, have students discuss high school possibilities in small groups or one large group. Advisors/counselors should maintain hard or digital copies of this exercise.

A note to the advisor/counselor: As students engage in this process, they may be questioning themselves about their qualifications for specific schools, particularly for any specialized high school that requires an admission test. This presents an opportunity for advisors and counselors to speak individually with students to help them fine-tune their lists of possible high schools. Particularly as students await admission decisions, individual meetings with students are advisable to help them in what may be a tense, anxiety-provoking time.

Advisors/counselors should also advise students to spend a day or part of a day at schools to which they have applied and been accepted. Particularly, private schools, both independent and Catholic, may offer this option to accepted students. While visiting an Open House at a school may provide much information, sitting in freshman classes and experiencing the school atmosphere on a personal basis may help students decide the next step in their K–12 journey. However, caution students and parents alike that these visits are not meant for the students to miss school on multiple days.

As advisors/counselors continue their work with eighth graders about high school options, it is essential to include parents in this process. While sample **Agendas of Parents Meetings on High School Options are available in Chapter 14,** we also suggest the following ways to include parents in this all-important process.

- Conduct a meeting for parents of students in seventh grade during second semester. At this meeting, administrators and counselors present high school options and required testing. Seek help from the district superintendent who may be able to assist individual middle schools in delineating high school options for students. If testing is required, inform parents of ways that students can prepare for entrance exams.
- In the following year, conduct a second meeting for parents of students who are now in eighth grade. At this meeting, administrators and counselors should ascertain that families and students understand their high school

options and testing information. Again, seek assistance from the district superintendent in presenting high school possibilities. Most importantly, this meeting is an ideal time to make sure that all students are on the right course for high school.

- Counselors may also conduct individual meetings with students and parents in both seventh and eighth grades. As in the Parents Meeting format, students and parents should be informed about high school options, any testing requirements, and suggestions for test preparation. Like the Parents Meeting format, these individual meetings are a great way to ensure that all eighth graders are finding their best high school possibility.

Eighth grade is a pivotal transition year for young adolescents. As they enter the last year of middle school, many students now find themselves examining life ahead of them: high school and the changes that it will bring. Not only are students thinking about how their day-to-day life may change, but also the new opportunities that their high school may offer. Middle schoolers are on the brink of a new life era. Helping them find the best high school match is truly important!

Helping Students and Parents in the Transition to High School

See **Chapter 15** for more information regarding assistance for students and parents as they transition to high school. **The second semester eighth-grade advisory/group counseling sessions are also presented in Chapter 15** as well as a sample agenda for a parent-student meeting on transitioning to high school.

A final note...

What to do during the summer? No matter where you live, enriching summer programs exist. Look for possibilities in your area. Again, using NYC as an example, examine some of the specialized programs and summer activities that facilitate development of skills and confidence. Here are just a few examples:

- **Summer Rising 2026** is a partnership between New York City Public Schools and the Department of Youth and Community Development (DYCD). Summer Rising is free and available to all NYC students in grades K–8, pending seat availability. Students in Summer Rising will have access to both academic and enrichment programming over the course of the day, including field trips, arts activities, and outdoor recreation.[16]
- **Oasis Day Camp** has multiple sites in New York City. Through hands-on experiences, collaborative projects, and exploration of nature, this camp encourages children to express themselves, build friendships, and embrace fun in a happy and dynamic setting.[17]
- **New York University** offers immersive academic summer experiences for students grades 5–12. Choose from more than 50 unique programs that fit a student's academic interests, goals, and schedule. NYU offers immersive academic experiences for students grades 5–12 during the summer and throughout the academic year, spanning everything from weekend workshops to one–two-week intensive programs through 6–12-week courses.[18]
- **InsideSchools New York City** offers children and teens a wide range of after-school and summer activities—from paid museum internships to free science research programs. The free and low-cost programs listed here are a great way to explore new interests, get extra support, and supplement what is being taught during the school day.[19]
- **New York Edge** is the largest provider of afterschool and summer camp programming in New York City and Long Island, which offers daily activities for more than 25,000 students across NYC's five boroughs. Projects range from arts to sports and wellness to college and career readiness.[20]

- Also check local churches and YMCA for summer activities.

Make summer vacation an exciting, growing experience!

Notes

1 "High School Choice Programs." *Go CPS, Chicago Public Schools.* https://www.cps.edu/gocps/high-school/explore/choice-programs/
2 "Specialty Schools and Programs." *Dallas Independent School District.* https://www.dallasisd.org/parents-students/enroll/specialty-schools-and-programs
3 "Zones of Choice." *LAUSD.* https://zoc.lausd.org/
4 "NYCPS Data at a Glance." *NYC Public Schools.* https://www.schools.nyc.gov/about-us/reports/nycps-data-at-a-glance
5 "NYC Public Schools: What Does It Mean to Be the Nation's Largest School District?" *NYU Steinhardt.* https://steinhardt.nyu.edu/research-alliance/research/spotlight-nyc-schools/nyc-public-schools-what-does-it-mean-be-nations
6 "Apply to High School." *NYC Public* Schools. https://www.schools.nyc.gov/enrollment/enroll-grade-by-grade/high-school
7 "NYC Common Online Charter School Application." *SchoolMint.* https://nyccharterschools.schoolmint.net/signin
8 "Specialized High Schools." *NYC Public Schools.* https://www.schools.nyc.gov/enrollment/enroll-grade-by-grade/specialized-high-schools
9 "Specialized High Schools." *NYC Public Schools.* https://www.schools.nyc.gov/enrollment/enroll-grade-by-grade/specialized-high-schools
10 "Specialized High Schools Admissions Test." *NYC Public Schools.* https://www.schools.nyc.gov/learning/testing/specialized-high-school-admissions-test
11 "Specialized High Schools." *NYC Public Schools.* https://www.schools.nyc.gov/enrollment/enroll-grade-by-grade/specialized-high-schools
12 Manhattan Hunter Science High School. https://www.mhshs.org/
13 Academy of American Studies. https://www.academyofamericanstudies.com/
14 Frank Sinatra School of the Arts. https://franksinatraschoolofthearts.org/index.jsp
15 New Explorations in Science, Technology & Math School. https://www.nestmk12.net/
16 "Summer Rising." *NYC Public Schools.* https://www.schools.nyc.gov/enrollment/summer-rising
17 Oasis Day Camp. https://oasischildren.com/
18 "High School and Middle School Programs." *NYU.* https://www.nyu.edu/admissions/high-school-and-middle-school-programs.html
19 Inside Schools. https://insideschools.org/resources/free-programs
20 New York Edge. https://newyorkedge.org/

PART III

Developing Critical Thinking Skills for Future Academic and Personal Success

The Role of Teachers, Counselors, and Administrators

11

Best Lesson Plans—Models that Develop Critical Thinking Skills

As students enter middle school, they are encountering multiple challenges in all areas. From meeting new classmates and teachers to using lockers and changing classes, the world of elementary school may seem in the distant past. Added to these differences is a much more demanding challenge: moving from a more factual learning to a more analytical synthesis that helps students develop higher critical thinking skills, an essential component in both personal and academic achievement. Once inside the middle school classroom, students will encounter academic challenges that will promote their development of these necessary higher level critical thinking abilities.

Many middle school students may have already engaged in moving from concrete thinking to a higher level of critical thinking. "Specifically, in fourth and fifth grades, these students need to demonstrate ...... critical thinking skills and apply self-motivation and self-direction to learning."[1] Now as sixth graders, students must engage in an ongoing expansion of critical thinking skills that are essential to personal and academic success.

First, let us look at the components of critical thinking:

- Observation: The ability to notice and predict opportunities, problems, and solutions.

DOI: 10.4324/9781003597292-14

- Analysis: The gathering, understanding, and interpreting of data and other information.
- Inference: Drawing conclusions based on relevant data, information, and personal knowledge and experience.
- Communication: Sharing and receiving information with others verbally, nonverbally, and in writing.
- Problem solving: The process of gathering, analyzing, and communicating information to identify and troubleshoot solutions.[2]

"In other words, good critical thinkers know how to analyze and evaluate information, breaking it down to separate fact from opinion. After a thorough analysis, they feel confident forming their own opinions on a subject."[3] Especially as young adolescents enter sixth grade, they are in a formative stage, ready to engage in critical thinking. Building upon this higher level of thinking helps middle school students to establish the foundational basis of strong study habits and skills that are fundamental to reaching the College and Career Readiness Standards.

*Another key resource is the ASCA Mindsets and Behaviors for Sixth and Seventh Grade Curriculum Planning Focus that demonstrates the importance of middle school students' critical thinking skill development.[4]

Lesson Plans That Develop Critical Thinking

In this chapter, we present examples of effective middle school lesson plans that require students to think in a more expanded fashion—ideally, outside of the box for a middle schooler as they move from factual learning to more critical thinking that involves synthesis and analysis. These lesson plans for sixth grade students demonstrate how teachers can effectively build upon students' interests in learning as they engage in an interactive approach that involves all in the learning process. The choice of these particular lesson plans from experienced teachers working with entering middle school students is a deliberate one: each lesson plan demands the involvement of all students in discussions and activities that work to develop effective critical thinking abilities. Also threaded into these lesson plans is an interactive learning approach that includes all students in the learning process. While these plans use SMARTboards and monitors, for schools not using this equipment, blackboards and/or bulletin boards may be substituted.

Lesson Plan 1 is designed as an introductory English Language Arts class which may be conducted over several days. The novel *Wonder* was required summer reading for these entering middle school students. While the teacher is relating the core values in the novel to her school's values, teachers may simply discuss the thematic values of the novel.

Our Core Values in Action: *An Exploration of How Values Are Revealed in Literature*

Kristen Lollis Bates, Fenn School, Concord, MA

This lesson incorporates the novel *Wonder*.[5]
Palacio, R. J. *Wonder*. Penguin Random House, 2012.

While this lesson is based on Wonder *and a specific school's core values, the concept and general overview can be applied using different values and a variety of middle school novels.*

Overview

Designed for 10–12 year old students, this lesson plan is a great way to build community at the start of the year in language arts. It is a multi-day lesson that explores how this independent school's core values intersect with the values revealed in *Wonder*. While Fenn's values are empathy, courage, respect, and honesty, the concepts and structure of the lesson can be adapted using different values.

The lesson focuses on collaboration, public speaking, and providing evidence to support a claim. As a culminating activity, students work in small groups to create a poster that demonstrates how their selected core value is an integral part of the book.

Learning Objectives

Students will be able to:

- Defend their claim that a particular value is the most prevalent in *Wonder.*
- Identify examples and provide quotes from the book to support their claim.
- Collaborate with group members to create a poster presentation that demonstrates their comprehension and understanding.

Materials and Resources

Wonder (enough copies for each student)
Highlighters, pens, pencils
Post-its (enough for each group)
Chart Paper
Each student's journal

Lesson Procedures

This lesson typically takes three days to a week.

A series of Google Slides accompanies the lesson, providing directions and visual cues for the students.

Slide Breakdown:

- **Slide 1**: Opening Quote—"Values are like fingerprints. Nobody's the same, but you leave them all over everything you do."—Elvis Presley

Students discuss the quote and share their thoughts about values.

- **Slide 2**: *What values are important to you or your family?* Students use post-its to list personal values, which they can later share while walking around the room.
- **Slides 3 and 4**: Visual cues prompting students to open their journals and set up a page to reflect on their values and class discussion (after partner conversations).
- **Slide 5**: *4 Corners Activity*
  Each corner is labeled with one of the school's core values. Students identify and discuss which values appear in *Wonder.*
- **Slide 6**: *Where and When Does This Value Show Up?* Students discuss and debate where these values emerge in the book. A character list is provided as a reference.
- **Slide 7**: Journal Setup—Students copy the prompt and write about how one value is important in *Wonder.*
- **Slide 8**: Group Project Instructions
  Includes a reminder about positive group collaboration.
  - **Step 1**: How does ________ show up in *Wonder*? Brainstorm examples on chart paper.
  - **Step 2**: Find quotes from the text that reflect this value. Include page numbers.
  - **Step 3**: Create a poster that illustrates your findings!

Day 1: Exploration of Personal Values

Journal Freewrite Prompt: *This can be a list or a personal reflection (can be adapted for each student's needs).*

Students reflect on:

- What values do you look for/appreciate in a friend?
- Why are these values important to you?
- What values are most important to your family?
- Do you share the same values? Why or why not?

Pair/Share and Link-Up

- Students write down 3–5 personal values on individual post-its.
- They trade post-its with a partner.
- In a circle, students share their partner's values.
- If others share the same values, they "link up" with those students.

Class Discussion

- As students group in different parts of the room, they share observations.
- Discuss: Are there overlaps between school and personal values?
- Why might overlap exist?

Homework: Journal Reflection

- When do these values show up in your life?
- How are we impacted when they are not in place?
- Think of real-life examples to share in class.

Day 2: The Intersection of School Values and *Wonder*

(Teachers may use honesty, respect, empathy, and courage, or substitute values relevant to their own school community.)

4 Corners Activity

- Each corner represents one of the core values.
- Students stand in the corner that they believe is the most prevalent in the book.

Brief Debate / Share-Out

- Students in each corner explain and defend their position.

Group Work

- In small groups, students discuss their value, search for quotes in *Wonder* that support their claim, and write them on chart paper.
- They must include page numbers and brief context for each quote.

Homework or In-Class Extension

- Each student selects one quote and explains in writing how it supports their claim.
- How does the quote show the importance of the value?
- What is the result when the value is present or absent?

Day 3 (+ Additional Days)

Group Project: Poster Creation and Presentations

- Students work in groups to create a poster.
- Using markers or paint pens, they illustrate quotes and examples from *Wonder*.
- The poster should clearly reflect their chosen value.

Class Presentations and Closure

Students present their posters, including:

- The selected value and examples from *Wonder*.
- Explanation of how the quotes support their claim.
- How the illustrations reflect the chosen value.
- Why this value is important to our classroom and school community.
- Reflection: What is the classroom like when this value is present? What happens when it's not?
- Which type of classroom would you prefer?

Assessment

Formative Assessment

- Read and comment on journal entries.
- Circulate during group work to observe contributions.

Summative Assessment

- Review each group's poster and presentation.
- As a post-reflection, students write about how they contributed to the group and the final presentation.

Reflection

This unit lays the foundation for community-building at the beginning of the year. It uses shared values and goals to guide expectations for collaboration and behavior. Students begin the year practicing how to find and use textual evidence and how to present their ideas in a group. Just as importantly, they build relationships and explore social–emotional learning that sets a positive tone for the rest of the year.

What Kristen Lollis-Bates presents in this multiday lesson plan is a thoroughly well-designed, thought-provoking analysis of *Wonder* that consistently engages her students. Sixth graders are actively involved in each step of this lesson plan as they choose a particular value that they believe is important, find quotes that support their choice of that value, and reflect upon the values presented in the novel. In essence, this activity promotes the solid development of critical thinking abilities and skills. Furthermore, this lesson plan incorporates several Journal Entry activities that are especially effective as students must choose a particular value and explain in writing why that value is important to them as individuals as well as to the overall community. From the Journal Freewrite to the Group Discussion to the Journal Reflection, students are engaged in the process of critical thinking and analysis of the values presented in *Wonder.* The four Corners and Brief Debate activities are interactive, ensuring that all students are participating in the analysis of the values presented in the book. Again, the closing activities of Day Three where students create posters reflecting their chosen value and make a group presentation to the class, effectively include all students in the learning process. As Ms. Lollis-Bates comments in her Reflection, this lesson plan includes community building while at the same time, students learn how to find and use textual evidence and to present their ideas to a group. Clearly, these young middle school students are developing their critical thinking abilities, skills that are based upon analysis and synthesis.

Lesson Plan 2 is designed for a Sixth Grade Social Studies/US History class with the intent of helping middle schoolers understand the importance of the events that occurred at Lexington and Concord and their relationship to the American Revolution. As they examine materials provided, students see the "why" behind colonial response in an interactive approach that involves all students.

Heather Kay, Utica Community Schools, MI

Lesson Plan: The Battles of Lexington and Concord

Grade Level: Sixth Grade
Subject: Social Studies / US History
Topic: The American Revolution Begins

Objective

Students will be able to:

- Describe the events of Lexington and Concord.
- Explain why these battles are called "the shot heard 'round the world."
- Understand the importance of ordinary people in starting the Revolution.

Materials

- Map of Massachusetts (showing Boston, Lexington, Concord).
- Copies of the poem (provided below) or project the poem on the board.
- Copies of the song (provided below).
- Smartboard.

Hook (5 minutes)

Ask: *"Imagine you and your neighbors had to stand up to one of the strongest armies in the world. Would you be brave enough to face them?"* Tell students: *"This happened in 1775, when farmers and townspeople faced British soldiers in Lexington and Concord near Boston."*

The Battles of Lexington and Concord—Article Review (Main Points)

Project on the Smartboard

The First Shots of the American Revolution

On the morning of April 19, 1775, two small towns in Massachusetts—Lexington and Concord—became very famous. Why? Because they were the places where the American Revolution began.

For years, the American colonists had argued with Great Britain about taxes and laws. The colonists wanted more freedom to make their own choices, but the British Parliament wanted to stay in control. Tensions were high, and many people believed that fighting might soon break out.

The British March

The British army in Boston heard that colonists were storing weapons in Concord. So, on the night of April 18, 1775, hundreds of British soldiers marched quietly toward the town. But the colonists were warned. Riders like Paul Revere spread the alarm by shouting, "The British are coming!"

Lexington: The First Shots

Early in the morning, about 70 colonial farmers and townspeople—called Minutemen—stood waiting on the village green in Lexington. The British soldiers ordered them to leave. Suddenly, a shot rang out. No one knows who fired it, but it started a battle. Several colonists were killed, and the soldiers marched on to Concord.

Concord: The Fight Grows

At Concord, more Minutemen gathered. They fought the British at the North Bridge and forced them to retreat. On the long road back to Boston, colonists hid behind trees and walls, firing at the soldiers the whole way.

Why It Matters

These battles were small, but they were very important. They proved that ordinary colonists were willing to fight for their rights. That first mysterious gunshot is often called "the shot heard 'round the world" because it began a revolution that would inspire people everywhere to fight for freedom.

Lesson Enhancements

Summary/discussion of the poem "Concord Hymn" by Ralph Waldo Emerson[6]

Emerson wrote the poem in 1837 to honor the Battle of Concord. The most famous line calls the first musket fire "the shot heard 'round the world" because it signaled the start of a revolution that inspired people far beyond America.

Poem (Original, kid-friendly, six lines, copyright friendly)

"The Shot Heard 'Round the World"

In Lexington the morning came,
The Redcoats marched to stake their claim.
The farmers stood, their voices strong,
Though no one knew who fired wrong.
From Concord's bridge the fight unfurled,
And freedom's cry shook all the world.

Lexington and Concord Song (to the tune of *Yankee Doodle*)—Students may learn this song to further their knowledge. They may sing in groups and perform the song for their peers.

The Redcoats marched from Boston town,
To take the guns and powder,
But riders cried, "The troops are near!"
The news spread even louder.

(Chorus)
Lexington and Concord too,
Minutemen stood ready,
They fired the "shot heard 'round the world,"
Their hands were strong and steady.

At Lexington the first shots rang,
No one knows who started,
The farmers faced the British line,
So brave and real tough-hearted.

(Chorus)
Lexington and Concord too,
Minutemen stood ready,
They fired the "shot heard 'round the world,"
Their hands were strong and steady.

After singing the song, students engage in the next step of this lesson plan: creating a poster that shows the routes that the British soldiers and the Minutemen traveled and where

the fighting took place during the Battles of Lexington and Concord.

Materials needed: posterboards, red, blue, and black magic markers.

- Project on the Smartboard an image of a map showing the locations of Boston, Lexington, and Concord.
- Break students into small groups where they create posters that map the three towns and show the movement of the British soldiers and the Minutemen.
- Each group presents their poster and explains the movement of the British soldiers and how the Minutemen responded.

Discussion Questions (5 minutes)

- Why do you think colonists were willing to fight the British?
- Why is it called "the shot heard 'round the world"?
- How were ordinary people important in this story?

Ms. Kay presents a multi-tiered approach to helping sixth-grade students understand the key importance of the Battles of Lexington and Concord in American history. An excellent feature of this lesson plan that promotes students' understanding of these events is that Ms. Kay uses an interdisciplinary approach to teaching the significance of these two battles, one that actively engages students in several ways. After presenting the *Hook,* she uses the Smartboard to present the key information that the class should learn. As the class progresses, students are discussing and analyzing the all-important Why behind these historical events. What is also an engaging feature of this lesson plan is that Ms. Kay incorporates both a song and a poem into her lesson plan. This interdisciplinary approach helps students to understand that historical events are not isolated occurrences from what is happening in their society. Having students read Emerson's "Concord Hymn" written in 1837 to celebrate this event, enables students to see how literature, in this instance poetry, is reflective of societal beliefs. As students then read and sing the lyrics of "The Shot Heard 'Round the World," they are increasingly engaged in not only understanding the historical events but also seeing how this popular song reflected the viewpoints of the people of that time. Furthermore, the use of the map once again actively involves the students in understanding these events in a hands-on method. The concluding discussion questions truly require students to demonstrate that they know the how and the why intrinsic to the Battles of Lexington and Concord. Clearly, these sixth-grade students are actively involved in critical thinking based upon analysis and synthesis.

Lesson Plan 3 is designed for a Sixth Grade Math class presenting Greatest Common Factor (GCF) and Least Common Multiple (LCM). Students will use prime factorization to find the factors of a number. Again, this lesson provides an interactive learning approach that includes all students.

Tony Andrus, Schwarzkoff Middle School, Sterling Heights, MI

Lesson Plan: GCF and LCM with Prime Factorization

Grade Level: Sixth Grade

Topic: Greatest Common Factor (GCF) and Least Common Multiple (LCM)

Time: 45–50 minutes

Learning Objectives

By the end of this lesson, students will be able to:

Use prime factorization to find factors of a number.

1. Use a Venn diagram to organize factors.
2. Find the GCF of two numbers.
3. Find the LCM of two numbers.

Materials

- Whiteboard or chart paper
- Markers
- Student notebooks
- Printed blank Venn diagrams (or students can draw them)

Vocabulary

- Prime number: A number with exactly two factors (1 and itself).
- Composite number: A number with more than two factors.
- Factor: A number that divides another number evenly.
- Multiple: The product of a number and an integer.
- GCF: The largest factor two or more numbers share.
- LCM : The smallest multiple two or more numbers share.

Direct Instruction (Teacher Modeling)

Step One: Prime Factorization

Example: Find the prime factorization of 36 and 60.

- $36 \rightarrow 2 \times 2 \times 3 \times 3$
- $60 \rightarrow 2 \times 2 \times 3 \times 5$

Step Two: Venn Diagram

- Draw two circles overlapping.
- Put factors of 36 in the left circle, 60 in the right circle, and shared prime factors in the middle.

Example

- Left (36 only): 3
- Middle (both): 2, 2, 3
- Right (60 only): 5

Step Three: Finding GCF and LCM

- GCF: Multiply the factors in the middle section → 2 × 2 × 3 = 12
- LCM: Multiply ALL factors from the Venn diagram → 2 × 2 × 3 × 3 × 5 = 180

Guided Practice (Students with Teacher Support)

Find the GCF and LCM of 18 and 24:

- 18 = 2 × 3 × 3
- 24 = 2 × 2 × 2 × 3
- Venn Diagram:
  - Shared: 2 × 3 GCF = 6
  - All together: 2 × 2 × 2 × 3 × 3 LCM = 72

Independent Practice

Students solve in pairs:

1. Find the GCF and LCM of 20 and 50.
2. Find the GCF and LCM of 12 and 15.

Wrap-Up

- Ask: "When do we use GCF in real life? When do we use LCM?"
  - GCF → dividing things into equal groups (like splitting snacks).
  - LCM → finding when things happen together (like traffic light cycles).
- Quick check: "What's the difference between a factor and a multiple?"

Additional:

Factor "Relay Race"

1. Write numbers (e.g., 18, 24, 12, 15) on the board.
2. Divide students into **teams**.
3. Each team must **run to a factor station** (sticky notes around the room labeled with prime factors like 2, 3, 5, 7, 11 etc.) and collect the correct prime factors for their number.
4. Once they've collected them, they lay them out in a Venn diagram drawn on the floor with tape or chalk.
5. First team to correctly build the diagram and identify GCF and LCM wins.

In this lesson plan, Mr. Andrus presents an interactive approach to teaching the GCF and LCM. The objectives of this class are that students achieve an understanding of how to use a Venn diagram to organize factors; to find the GCF of two numbers; and to find the LCM of two numbers. The instruction of the class accomplishes this learning in a step-by-step process where students learn the mathematical vocabulary needed first by presentation and examples and then by doing the actual work involved in the Guided Practice. In the Direct Instruction, Mr. Andrus presents each step and illustrates how to accomplish that math work through an example. As students move into the Guided Practice, they are given related problems and work independently with the assistance of the teacher as needed to solve the assigned problems. The next step of achieving the aims of the lesson plan is that students work in pairs to solve assigned math problems. Mr. Andrus then completes a Wrap-Up reinforcing in the minds of these sixth graders what they have learned in this class. Again, these young middle schoolers are engaged in critical thinking throughout this class. The Factor Relay Race is a wonderful finale as students actively participate in a team effort to collect the correct prime factor sticky notes so that they may create a Venn diagram on the floor with tape or chalk. This lesson plan is an excellent example of helping middle school students learn essential mathematical vocabulary/terms and how to use problem solving to create an accurate Venn diagram. As students work individually, in pairs, and on teams, they are actively engaging in the critical thinking process.

Lesson Plan 4 presents a review of forms of energy that sixth graders have been studying in science class. This layered interactive approach to review helps students to understand what they have learned and the importance of review to assist them in achieving academic success.

Heather Kay, Utica School District, MI

Interactive Lesson Plan: Forms of Energy (Review)

Grade Level: 6th

Time: 45–50 minutes

Learning Objectives

Students will:

1. Review the main forms of energy (mechanical, thermal, electrical, chemical, light, sound, nuclear).
2. Identify real-life examples of each energy type.
3. Show understanding through movement-based and group activities.

Materials

- Chart paper or whiteboard.
- Markers.
- Energy cards (examples of energy forms written on slips of paper—e.g., "a guitar string vibrating," "a flashlight," "a microwave," "burning wood," "a rollercoaster at the top of a hill").

- Small sticky notes.
- Optional: projector/slides with images of energy examples.

Lesson Flow

1. **Hook / Warm-Up (5 minutes)**
   - Teacher asks:
     "What are some forms of energy you've learned about this year?"
   - Students brainstorm aloud. The teacher writes answers on the board.
   - Quick reminder: "Energy is the ability to cause change or do work. It comes in many forms!"
2. **Energy Charades (10 minutes)**
   - Teacher gives one student an *Energy Card* (e.g., "toaster," "drums," "campfire").
   - Student acts it out *without speaking*. Class guesses via shout-outs.
   - Keep score for fun, or just rotate quickly through volunteers.
3. **Energy Relay Game (10 minutes)**
   - Split class into 2–3 teams.
   - Teacher asks an energy question or shows a picture.
   - First student in line runs to the board and writes the correct energy type (mechanical, chemical, etc.).
   - They tag the next person, and so on.
   - Fastest team with the most correct answers wins.
4. **Wrap-Up / Exit Ticket (5 minutes)**
   - Each student gets a sticky note and writes:
     1. Their favorite form of energy
     2. One real-life example of it
   - Stick on the board under the correct energy category.

Assessment

- Observation during charades and stations.
- Accuracy in relay answers.
- Exit tickets for individual check.

Energy Cards—Forms of Energy Review

Cut out these cards for games like Charades, Stations, or Relay. One side shows the example, and students guess the form(s) of energy. The teacher version includes answers.

Example: Roller coaster at the top of a hill

Answer: Mechanical (potential)

Example: Burning wood in a campfire

Answer: Chemical → Thermal and Light

Example: A flashlight shining

Answer: Electrical → Light

Example: A microwave heating food

Answer: Electrical → Thermal

Example: A guitar string vibrating

Answer: Mechanical → Sound

Example: The Sun shining on Earth

Answer: Nuclear → Light and Thermal

Example: Lightning bolt in a storm

Answer: Electrical

Example: Food giving you energy to play soccer

Answer: Chemical → Mechanical

Example: Plugging in a phone charger

Answer: Electrical

Example: Drums being hit

Answer: Mechanical → Sound

Example: Ice melting

Answer: Thermal

Example: Solar panels making electricity

Answer: Light → Electrical

Example: Fireworks exploding

Answer: Chemical → Sound, Light, Thermal

Example: Wind turning a turbine

Answer: Mechanical → Electrical

In this lesson plan, Ms. Kay guides students through a solid review of the forms of energy. As students identify the various forms of energy, they must also relate these forms to real-life examples of each energy type. This lesson plan achieves an effective review while at the same time engaging students in an interactive overview of energy forms and their uses. The Energy Charades using the *Energy Cards* involves the whole class as students act out energy examples and students shout out the energy form being depicted—just as people play charades. The Energy Relay Game further continues the review of the energy forms and examples as the teacher divides the class into teams. As the teacher asks a question or shows a picture, students race to the board to write the correct energy type. Once again, all students are actively involved in this review of energy forms and applications. The Exit Ticket completes the class review where students write out their favorite form of energy and a real-life example of its use. This lesson plan for a systematic review of energy forms and examples of their uses differs quite dramatically from the traditional notes on the board style of review. In this more interactive approach, sixth grade students are reviewing all the forms of energy and examples of their use in an interactive manner that engages them; they understand the content and enthusiastically engage in the games from the Energy Charades to the Relay Race to the Exit Ticket. As Ms. Kay has intended in this lesson plan, all students were actively involved participants in this energy forms and uses review.

Many thanks to the teachers who shared these excellent lesson plans that promote critical thinking in the all-important middle school years. Each of these lesson plans shared a common feature: active interaction of students in the learning process. Sixth grade students are not passive learners. Rather, they thrive on the "doing activities" that each lesson plan incorporated. This interactive learning promotes critical thinking analysis and synthesis as students reflect on what they are learning and use specific examples as evidence to support their conclusions. From the values of *Wonder* to the commitment of the Minutemen to GCF and the LCF to energy forms and their uses, these lesson plans demonstrate how to engage middle school students in an interactive manner that develops their critical thinking abilities. Teachers, well done!

Notes

1 Curry, Jennifer R. and Amy Milsom. *Career and College Readiness Counseling in P-12 Schools,* 2nd edition. New York: Springer Publishing Co., 2017, pp. 185–186.

2 "What is Critical Thinking and Why Do We Need To Teach it?" *We Are Teachers,* January 2023. https://www.weareteachers.com/what-is-critical-thinking/

3 "What is Critical Thinking and Why Do We Need To Teach it?" *We Are Teachers,* January 2023. https://www.weareteachers.com/what-is-critical-thinking/

4 ASCA Mindsets and Behaviors for Sixth- and Seventh Grade Curriculum Planning Focus* Curry, J. R. and A. Millson. *Career and College Readiness Counseling in P-12 Schools,* 2nd edition. Spring Publications, 2017, p. 219 and 239.

5 Palacio, R. J. *Wonder.* Penguin Random House, 2012.

6 Emerson, Ralph Waldo. "The Shot Heard Round the World." https://www.pbs.org/wgbh/americanexperience/features/patriotsday-shot-heard-round-world/

12

How Standards and Testing Can Help Middle School Teachers, Counselors, and Administrators

Education in the United States has historically been in the domain of the state governments. However, as we entered into the 21st century, movements have developed to ensure that the education of students nationwide is meeting common standards, particularly in English language and mathematics, standards that will prepare them for the postsecondary world. These standards recognize that students are on a K–12 journey, one that will prepare them for career and college readiness. The movement to create national standards includes many groups and organizations that share the belief that all students should have an education that will prepare them for adult life in a fast-growing, global society. Both the Common Core State Standards (CCSS) and the College and Career Readiness Standards (CCRS) grew out of this national movement aimed at ensuring that all students are engaged in an education that will prepare them for their postsecondary endeavors.

DOI: 10.4324/9781003597292-15

A Brief History

In 2004, Achieve, the Thomas B. Fordham Foundation, and the Education Trust released a report that encouraged states to align K–12 standards with college and workplace expectations so that high school would prepare students for the real world. This report sparked work by states to recalibrate their standards in accordance with higher education and employer expectations. Later in 2008, college and career readiness became the goal for all students, resulting in the impetus for the Common Core State Standards initiative, which was established through a collaborative effort, primarily spearheaded by the National Governors Association Center for Best Practices and the Council of Chief State School Officers. These organizations initiated the Common Core State Standards, which are explicitly linked to college and career preparation. Additionally, organizations like the ACT and Achieve have also played significant roles in defining and promoting these standards. [1]The Common Core State Standards (CCSS) were developed to provide a consistent framework for what students should know and be able to do in English language arts and mathematics. This framework is designed to ensure that students are adequately prepared for college and careers.

According to the CCSS Initiative, "the Common Core State Standards build on the best of what states already had in place. They provide clear, consistent learning goals designed to prepare students for success in college, careers, and life. These standards clearly show what students are expected to learn at each grade level, so parents and teachers can better support student progress." Furthermore, the Common Core State Standards share the following features:

Key Features of the Standards:

1. Based on research and evidence.
2. Clear, understandable, and consistent.
3. Aligned with college and career expectations.
4. Focused on rigorous content and the application of knowledge through critical thinking skills.

5. Built on the strengths and lessons from current state standards.
6. Informed by top-performing countries to prepare students for success in a global economy.

According to the CCSS Initiative, research shows that mastering these standards is crucial for students to thrive in today's global economy, whether they pursue college or a career.[2]

Currently, forty-six states and the District of Columbia have voluntarily adopted the Common Core State Standards (CCSS) and have grade learning expectations for students in grades K–12 for Mathematics and for English Language Arts and Literacy.

> While states have had standards for more than 15 years, the CCSS set clear, consistent and high learning goals focused specifically on ensuring that students are college- and career-ready.....In addition, to the Common Core Standards, eighteen states and the District of Columbia have adopted the Next Generation Science Standards (NGSS) which are K–12 science content standards. Similar to the CCSS, the NGSS was developed by states to improve science education for all students by introducing a set of research-based, up-to-date K–12 science standards that were consistent across states.

Expanding the standards to include science content is a solid movement to ensure students' college and career readiness.[3]

In addition to the CCSS, the College and Career Readiness Standards (CCRS) are high-quality benchmarks aimed at ensuring that every student has the skills to compete in today's global, knowledge-based economy. These standards "represent goals for what students should know and be able to do while learning academic content. They differ from curriculum, which provides educators with an outline of what should be taught in the classroom."[4] Another important aspect of the CCRS is the recognition that these benchmarks are achieved over the ongoing years from K–12, that is from kindergarten/elementary to middle school to high school. In the middle school years, the standards focus on building the foundational skills and habits that prepare students for future academic and professional success. Here, it is important to note that: "No single entity created the "College and Career Readiness Standards" as different organizations and state and national initiatives have developed their own sets of standards, often with input from educators, business leaders, and government bodies. Key players include the National Association of Colleges and Employers (NACE), the School Superintendents Association (AASA)'s "Redefining Readiness initiative and the ACT organization which developed its own set of standards based on research."[5]

Together, the CCSS and the CCRS strive to ensure that all students are engaged in a learning process that will prepare them for their postsecondary lives in a fast-growing, global economy.

The Role of the Federal Government in Education in Recent History

In 1965, as part of the Great Society legislation under President Lyndon B. Johnson, the Elementary and Secondary Education Act (ESEA) was passed. ESEA created a clear role for the federal government in K–12 policy, offering more than $1 billion a year in aid under its first statutory section, known as Title I, to districts to help cover the cost of educating disadvantaged students. The law has been reauthorized and changed more than half a dozen times since that initial legislation.[6] Later, in 2002, Congress passed the No Child Left Behind Act. Under the NCLB, the role of the federal government changed as this law allowed increased federal regulation of state school systems.

> The NCLB law—which grew out of concern that the American education system was no longer internationally competitive—significantly increased the federal role in holding schools responsible for the academic progress of all students. The NCLB law put a special focus on

> ensuring that states and schools boost the performance of certain groups of students, such as English-language learners, students in special education, and poor and minority children, whose achievement, on average, trails their peers. States did not have to comply with the new requirements, but if they didn't, they risked losing federal Title I money.

Losing funds clearly motivated states to comply with the new federal guidelines.[7]

In 2015, under President Barack Obama, the Every Student Succeeds Act was passed, replacing the NCLB Act. ESSA requires that student performance targets and school ratings are based on multiple measures, not merely achievement in test scores. While the No Child Left Behind (NCLB) Act represented a significant step forward for our nation's children in many respects, particularly as it focused on where students were making progress and where they needed additional support, regardless of race, income, location, disability, home language, or background, ESSA advanced equity by upholding critical protections for America's disadvantaged and high-need students. Also of essential importance is that ESSA requires that all students in America be taught to high academic standards that will prepare them to succeed in college and careers.[8]

In addition, ESSA requires that all schools measure non-academic outcomes, such as school climate and social-emotional learning, to assess student learning. Non-academic outcomes include such measures as:

- Student and educator engagement.
- Postsecondary readiness.
- School climate and safety.
- Student self-control and conscientiousness.
- Parent satisfaction and engagement.
- Rates of in-school and out-of-school suspensions and expulsions.
- Chronic absenteeism.

- Incidents of violence, bullying, and harassment.
- Graduation rates.
- Social-emotional learning.[9]

Thus, ESSA made significant inroads in requiring not only high academic standards but also in examining school climate on multiple levels. In particular, the emphasis on school climate and social and emotional learning promotes an educational atmosphere where students are actively engaged. This law also addresses issues such as chronic absence, bullying, and violence among other negative occurrences in the school environment, truly making ESSA a truly forward step in federal involvement in education.

Under President Joe Biden, as part of the American Rescue Plan of 2021, the Elementary and Secondary School Emergency Relief Act, also known as ESSER, was passed. This key act provides federal funds to help schools recover from the Covid pandemic period. Districts across the country have used ESSER funding to help create more collaborative classrooms by investing in the latest technology to foster an interactive learning environment for their students. Currently, ESSER funding has been extended to March 2026. Furthermore, as noted in Chapter 7, under President Joe Biden, the Infrastructure Investment and Jobs Act of 2021 has provided a historic $65 billion investment to expand internet access and use throughout the United States. By November 2023, all 50 states and 6 territories had submitted their Broadband Equity, Access and Deployment Program Initial Proposals.[10] However, as former Education Secretary Miguel Cardona has warned, a growing digital divide is occurring in the United States. In response to the widening digital divide, the 2024 National Educational Technology Plan presented a forward-thinking approach to reframing and realizing the potential of educational technology to enhance the instructional core, reduce achievement gaps, and improve student learning in our schools.[11] With federal funding currently in question and the dismantling of the Department of Education under President Trump, the digital divide will likely broaden. (See Chapter 7 for more detail.)

The PreACT 8/9 and PSAT 8/9 Tests: Methods of Evaluating Student Performance and School Curriculum

A Look at the PreACT 8/9

As middle school students journey through grades 6-8, how can standardized tests help administrators, teachers, and counselors assess both student and school success in developing the solid framework of knowledge, skills, and abilities essential for college and career readiness? Students' scores track not only individual progress, but also include areas of the school curriculum that may need further enhancement to promote student success.

The PreACT assessments are anchored in the College and Career Readiness Standards. Analyzing student performance in the four test areas—English, Mathematics, Reading, and Science—offers valuable insight into areas where curriculum may be targeted for improvement and opportunities for student growth. In addition, taking the PreACT offers students an early experience with the ACT test items and helps them become acquainted with the many useful resources on the ACT website that will provide valuable assistance on their journey to College and Career Readiness. (Please note the ACT allows students as young as eleven to create an account. Students participating in the advisory/group guidance systems have ACT accounts for career and early college searches.)

ACT College and Career Readiness Standards[12]

The ACT College and Career Readiness Standards® are the backbone of ACT assessments. The standards are empirically derived descriptions of the essential skills and knowledge students need to become ready for college and career, giving clear meaning to test scores and serving as a link between what students have learned and what they are ready to learn next. Parents, teachers, counselors, and students use the standards to:

- Communicate widely shared learning goals and expectations.

- Relate test scores to the skills needed in high school and beyond.
- Understand the increasing complexity of skills needed across the score ranges in English, mathematics, reading, science, and writing.
- The standards encompass the many paths available to students after high school, and they reflect our ability to provide insights related to both college and career readiness.

Approximately, every three to five years, the ACT National Curriculum Survey is conducted. This nationwide survey of educational practices and expectations collects data about what entering college students should know and be able to do to be ready for college-level coursework in English, math, reading, and science.

What is the purpose of this survey?

- The results of the survey are used to inform ongoing efforts to develop, refine, and update common academic standards such as the Common Core State Standards, as well as to inform policymakers and educators.
- The results are also used to help guide development of ACT's curriculum-based assessments and ensure that they meet the needs of college and career readiness.[13]

The ACT is committed to constant monitoring to ensure that ACT assessments meet both content and predictive validity standards. The ACT "continually uses research and performance results to inform the changes we will make to test blueprints, the ACT College and Career Readiness Standards, and the ACT College Readiness Benchmarks." The ACT's ongoing research assures that the four content areas tested—English, math, reading, and science—meet both content and college and career readiness standards.[14]

Administering the PreACT 8/9 also provides teachers, counselors and administrators with a valuable way of seeing not only student progress but areas where curriculum may need further development. When students take the PreACT 8/9, the

ACT provides schools with the **Student Growth Percentile (SGP) model** that shows how well a student performed relative to peers with similar score histories. SGPs provide a normative measure of growth, describing growth relative to other students.[15]

How the SGP model can assist teachers, counselors and administrators

SGPs can be used to address important questions such as:

- How much growth is needed to become on target for college and career readiness?
- How did my student's growth compare to other students nationally?
- How much does growth in my school vary across subject areas, grade levels, and student subgroups?
- Has growth at my school improved over time?[16]

Resources for SGPs are provided to schools participating in the PreACT 8/9 test. Included is the **SGP look up table** that measures growth across grade levels and assessments for each section of the test. This information may be used not only to measure student progress, but also to examine curriculum in the four areas tested: English, Mathematics, Reading, and Science.

Next, a Look at the PSAT 8/9

Likewise, the College Board's SAT Suite of Assessments assists both students and educators in evaluating student progress toward college and career readiness. For middle school students, the PSAT 8/9 provides important information about student progress. The College Board has recently revised the College and Career Readiness Benchmarks "Each assessment in the SAT Suite has an associated set of metrics called the college and career readiness benchmarks. The new college and career readiness benchmarks are based on actual student success in entry-level college courses."[17] Using these benchmarks are of great assistance as educators evaluate not only eighth grade middle schoolers' progress but also determine areas where curriculum may need further enhancement.

How the College Board's College and Career Readiness Benchmarks Can Assist Students and Educators

The SAT Suite of Assessments works together to help students and educators assess student progress toward college readiness from year to year. Benchmarks help:

- Identify students who are thriving and require greater challenges.
- Identify students who require additional academic support.
- Inform instructional and curricular enhancements throughout an institution.[18]

What are the SAT College and Career Readiness Benchmarks?

When examining the recently revised benchmarks, it is essential to remember that college readiness is a continuum—students scoring below the SAT benchmarks can still be successful in college, especially with additional preparation and perseverance. Students are considered college and career ready when their SAT section scores meet both the Reading and Writing and Math benchmarks.

> Students with an SAT Reading and Writing section score that meets or exceeds the benchmark have a 75% chance of earning at least a C in first-semester, credit-bearing college courses in history, literature, social sciences, or writing classes.
>
> Students with an SAT Math section score that meets or exceeds the benchmark have a 75% chance of earning at least a C in first-semester, credit-bearing college courses in algebra, statistics, precalculus, or calculus.[19]

These benchmarks are key targets for high school students and can provide crucial insight into both student progress and areas where the curriculum may need further strengthening. According to the College Board's *Educator's Brief*, "Moving to a model where students can more easily understand whether they are ready for certain types of college courses—and what to do if they are not ready—ensures assessments are not a stopping point but a look toward the future, offering action steps that students can take to strengthen their college readiness skills."[20]

How can These Benchmarks Assist Middle School Educators?

Middle school students who take the PSAT 8/9 see grade-level benchmarks on their score reports. Grade-level benchmarks represent the section scores on each assessment that students should meet or exceed to be considered on track for college and career readiness. These benchmarks are based on expected student growth toward SAT College and Career Readiness benchmarks at each grade. Student performance is compared to the College and Career Readiness benchmark, regardless of their current grade.[21]

CHART 12.1 8TH GRADE BENCHMARKS (PSAT 8/9)

- Reading and Writing: 390
- Math: 430

8th Grade Section Score Ranges

120–720 Point Scale

	Not Yet Approaching	Approaching	Meets/ Exceeds
Reading and Writing	120–360	370–380	390–720
Math	120–400	410–420	430–720

The College Board uses a color-coded system to indicate student performance relative to benchmarks.

- Green indicates scores that meet or exceed the benchmark.
- Yellow indicates scores that are approaching the benchmark.
- Red indicates scores that are significantly below the benchmark, requiring skill strengthening.[22]

How Educators Can Use Student Performance on the PSAT 8/9 to Improve the Middle School Academic Experience

In examining student scores, educators must keep two goals in mind: how to assist all students in improving their scores and how the curriculum may be enhanced to improve student growth in particular subject areas.

- Students who are on target for the SAT benchmarks may be encouraged to take more challenging classes.
- For students whose scores show that they may need more help, educators should connect them to classroom work that is focused on areas that need improvement.
- Educators should develop methods of providing additional assistance. Afterschool sessions in subject areas are an excellent way to give students the added explanations that they may need.
- Educators' analysis of student performance in the SAT subject areas may provide valuable insight into areas of the curriculum that may need further development.[23]

Both the ACT College and Career Readiness Standards and the College Board's College and Career Benchmarks are intended to help all students for life after high school. For an eighth grader taking the PreACT 8/9 or the PSAT 8/9, the postsecondary world may seem light years away; however, the reality is that high school commencement is not that far away. Analyzing the test results of students taking the PreACT 8/9 or the PSAT 8/9 may provide invaluable insight to educators as they strive to help all students be prepared for life after high school.

A Final Note...

Both the ACT and the College Board provide multiple resources for those participating in their programs—educators, counselors, parents and students alike—may find these online resources invaluable as they navigate the best options on their journey to postsecondary life.

A special note for middle school educators: another valuable benefit of the ACT is that students as young as eleven years old may establish ACT accounts. Once on the website, students may explore careers and colleges. As explained in Chapter 8, students take several inventories that lead them to a Career Map where they delve into the world of possible career paths ahead of them. Chapter 9 follows with initial research of colleges using the ACT website.

Of particular interest for educators are:

- ACT Resources- K–12 Administrators | ACT
  Help your students achieve college and career readiness with solutions, research, insights, videos, webinars, downloadable posters and brochures, and professional development.[24]
- ACT Resources- ACT Growth Modeling Resources
  Growth modeling resources can be used to help monitor the progress of students/schools and describe how well a student performed relative to peers with similar score histories.[25]

For further information and resources, see Appendix A.

The College Board also offers many resources to assist educators, counselors, parents and students. Once students are age 13, they may establish an account with the SAT. Even though they have not yet taken an exam, they may now take the College Board's Career Quiz and explore possible careers and colleges. For those students taking the PSAT 8/9, educators, counselors, parents and students can see their progress in reaching the College and Career Benchmarks of the SAT.

Resources of particular interest for educators are:

- Educator's Brief On Benchmarks and the SAT Suite[26]
- SAT® Score Relationships with College GPA: First-Year through Fourth-Year Cumulative GPA[27]

For further resources and information, see Appendix B.

Notes

1 "Getting Serious About College and Career Readiness." *Education Strategy Group*. https://edstrategy.org/resource/getting-serious-about-college-and-career-readiness/
2 Common Core State Standards Initiative. https://www.thecorestandards.org/read-the-standards/
3 "College- and Career- Readiness Standards." *Teachers College Columbia University*. https://sri.tc.columbia.edu/our-framework/college-and-career-readiness-standards/
4 "College- and Career- Readiness Standards." *Teachers College Columbia University*. https://sri.tc.columbia.edu/our-framework/college-and-career-readiness-standards/
5 "College- and Career- Readiness Standards." *Teachers College Columbia University*. https://sri.tc.columbia.edu/our-framework/college-and-career-readiness-standards/
6 Klein, Alyson. "No Child Left Behind: An Overview." *Education Week*, April 11, 2015. Policy & Politics, Every Student Succeeds Act. www.edweek.org. https://www.edweek.org/policy-politics/no-child-left-behind-an-overview/2015/04
7 Klein, Alyson. "No Child Left Behind: An Overview." *Education Week*, April 11, 2015. Policy & Politics, Every Student Succeeds Act. www.edweek.org. https://www.edweek.org/policy-politics/no-child-left-behind-an-overview/2015/04
8 "Every Student Succeeds Act (ESSA)." *US Department of Education*. https://www.ed.gov/laws-and-policy/laws-preschool-grade-12-education/every-student-succeeds-act-essa
9 "College- and Career- Readiness Standards." *Teachers College Columbia University*. https://sri.tc.columbia.edu/our-framework/college-and-career-readiness-standards/
10 Broadband USA. https://broadbandusa.ntia.gov/
11 "2024 National Education Technology Plan." *American Federation of School Administrators*, January 2024. https://www.theschoolleader.org/news/2024-national-education-technology-plan

12 "Practice Taking the ACT | PreACT | K12 Solutions." *ACT.* https://www.act.org/content/act/en/products-and-services/preact.html
13 "National Curriculum Survey." *ACT.* https://www.act.org/content/act/en/research/reports/act-publications/national-curriculum-survey.html
14 "ACT Growth Modeling Resources." *ACT.* https://www.act.org/content/act/en/research/services-and-resources/act-growth-modeling-resources.html
15 "ACT Growth Modeling Resources." *ACT.* https://www.act.org/content/act/en/research/services-and-resources/act-growth-modeling-resources.html
16 "ACT Growth Modeling Resources." *ACT.* https://www.act.org/content/act/en/research/services-and-resources/act-growth-modeling-resources.html
17 *K-12 Educator Brief: The College and Career Readiness Benchmarks for the SAT Suite of Assessments.* College Board, P.p. 1.
18 "Benchmarks- SAT Suite." *College Board.* https://satsuite.collegeboard.org/k12-educators/about/understand-scores-benchmarks/benchmarks
19 "Benchmarks- SAT Suite." *College Board.* https://satsuite.collegeboard.org/k12-educators/about/understand-scores-benchmarks/benchmarks
20 "Benchmarks- SAT Suite." *College Board.* https://satsuite.collegeboard.org/k12-educators/about/understand-scores-benchmarks/benchmarks
21 "Benchmarks- SAT Suite." *College Board.* https://satsuite.collegeboard.org/k12-educators/about/understand-scores-benchmarks/benchmarks
22 "Benchmarks- SAT Suite." *College Board.* https://satsuite.collegeboard.org/k12-educators/about/understand-scores-benchmarks/benchmarks
23 "Benchmarks- SAT Suite." *College Board.* https://satsuite.collegeboard.org/k12-educators/about/understand-scores-benchmarks/benchmarks
24 "ACT Resources- K-12 Administrators." *ACT.* https://www.act.org/content/act/en/k12-educators-and-administrators/administrator-resources.html
25 "Services and Resources Research." *ACT.* https://www.act.org/content/act/en/research/services-and-resources.html
26 *K-12 Educator Brief: The College and Career Readiness Benchmarks for the SAT Suite of Assessments.* College Board.
27 Marini, Jessica P., et al. *SAT Score Relationships with College GPA: First-Year through Fourth-Year Cumulative GPA.* College Board, October 2024.

13

The New Focus on Middle School

Why Partnerships and Grants are Important to College and Career Readiness for Middle Schoolers

Colleges, universities, and foundations have recently turned their attention to middle school as the critical age to begin development and motivation for college readiness, stronger academic skills, exposure to careers and the importance of education in adolescent development. Specifically, this chapter explores the role that college/university, foundation, and fund partnerships should have in the development of a successful middle school program. Essential to these partnerships is the active participation of counselors, teachers, school psychologists, and social workers in developing these relationships to further foster the impact of a middle school career and college readiness curriculum for student success.

College/University Partnerships with K–12 schools, specifically including 6–12 schools, and foundations may work with middle school children to enhance their college-going beliefs. Further in this chapter, we describe a highly successful multi-year partnership with a middle–high school in the Bronx borough of New York City. We as coauthors have experienced the highly effective benefits of such a partnership. The excitement of middle school students exploring possible future careers and colleges has only reinforced our belief about the importance of

DOI: 10.4324/9781003597292-16

assisting schools in their continuous development of students in the context of today's educational climate.

The ASCA National Model: Calling Counselors to Be Leaders in Educational Reform

For years, the American School Counseling Association (ASCA) has studied and developed models for the standardization of the practice of school counseling. Currently, ASCA (2025) has presented newly revised Professional Standards & Competencies that

> outline the mindsets and behaviors school counselors need to meet the rigorous demands of the school counseling profession and the needs of pre-K–12 students. These standards and competencies help ensure new and experienced school counselors are equipped to establish, maintain and enhance a school counseling program addressing academic achievement, career planning and social/emotional development.[1]

Of particular importance is that the role of the school counselor is not only responsible for maintaining and enhancing the school program but also, more importantly, the academic achievement, career planning and social–emotional development of students. The chapters of this book are designed to assist all educators, particularly counselors, as they address developmental plans for students in all three areas: academic achievement, career planning, and social–emotional development.

The standards movement has provided ASCA with a timely opportunity to better define the role of school counseling in the American educational system and establish similar goals, expectations, support systems, and experiences for all students across the country. Additionally, on the local level, National Standards offer an opportunity for school counselors, school administrators, faculty, parents, businesses, and the community to engage in conversations about expectations for students' academic success and the role of counseling programs in enhancing student learning.

The ASCA National Standards for Students also serves as the foundation for the ASCA National Model: A Framework for School Counseling Programs. Representing more than 50 years of research, theory, and practice, the National Model helps school counselors design and implement programs that meet the National Standards and establish school counseling as an integral component of the academic mission of their schools.[2]

The ASCA School Professional Standards and Competencies also expects that counselors practice within the ethical principles of school counseling according to the ASCA Ethical Standards. In addition, school counselors must adhere to whatever legal responsibilities that are a part of their role, including, among others, the unique legal and ethical principles of working with some possible students in a school setting. Counselors normally may also be ready to understand the potential issues regarding legal obligations, to adhere, for example, to the ethical and statutory limits of confidentiality. Finally, especially within the middle school, the counselor should fulfill legal and ethical obligations to families, teachers, administrators, and the school staff.[3]

Most importantly, the ASCA National Standards call for counselors to be actively involved in promoting the overall development of students. Active participation in partnerships with college and universities is an effective way to enhance the development of middle through high school students.

The American Psychological Association (APA) has suggested that interested researchers in developing partnerships should explore the following areas:

- **Educational Partnerships:** Look for studies that examine the benefits and challenges of partnerships between colleges and middle schools, focusing on how these partnerships can enhance student success and improve educational outcomes.The APA has partnered with schools interested in developing these partnerships. Dr. Fitzpatrick, as a New York State licensed psychologist and mental health counselor, has been very involved over time with local middle schools in the Bronx borough in NYC. The current concern is that the Department of Education continues to cut federal education research funding. APA and the American Counseling Association continue to foster partnerships. The APA Senior Director of pre-college and undergraduate psychology has stated: "Given the significant loss of federal dollars, reaching out to partner with schools is an important consideration in order to continue this area of research." Beth Schwartz, Ph.D., APA senior director of pre-college and undergraduate psychology education.[4]

- **Family/School Community Partnerships/School Ready Research:** Explore research that supports the idea that positive family/school community partnerships during middle school can enhance student success and the impact of such partnerships on student outcomes. School–Ready Partnerships are particularly important as middle school students must reach higher levels in their academic development in all areas of study.
- **Integrated Partnerships in Education:** Search for studies that discuss how integrated partnerships can help states and school districts build and sustain educational systems that address policy, practice, evidence generation, and implementation support.

These resources provide a foundation for understanding the dynamics and outcomes of partnerships between colleges/foundations, K–12 and middle schools (6–8), as well as the broader implications for educational reform and student success. As a member of the APA, Dr. Fitzpatrick has used these sources as an effective means of searching for partners in schools.[5]

How K–12 Schools and College/University Partnerships Help All Involved

K–12 College/University Partnerships: A New Look

College/University Partnerships can provide a positive influence on college and career readiness. This chapter engages its reader to understand the development of School–College/University Partnerships and Grants. Some colleges choose to partner only with high schools while others partner with K–12 schools. The overall goal to be described here in this book focuses on the connection between middle school environments and colleges connecting school counselors, teachers, social workers, and psychologists. It is important to understand that college counseling faculty are not the experts and K–12 professional educators are not the learners.

In 2000, Dr. Fitzpatrick applied for and was awarded a federal grant, titled the TITAN grant that awarded one million dollars over four years to Manhattan University and partner schools. Over the following years, she continued to receive increased funding to further the development with partner schools. With some of this funding, Dr. Fitzpatrick created and funded the Center for Optimal College Readiness at Manhattan University that serves K–12 children in partner schools in the greater New York City metropolitan area. Through these grants, Dr. Fitzpatrick placed graduate students in school counseling and special education to assist middle school children through the grant partnerships. Also included were partnerships not only with New York Public Schools, but also with schools in Yonkers and Mt. Vernon, north of the city. Furthermore, Dr. Fitzpatrick also gained funding from GEAR UP, Kids2College, and My Brother's Keeper, all of which focus on college readiness for students who may have never considered themselves capable or worthy of attending college. This money enabled multiple opportunities for middle school students to explore college possibilities.

The 6–12 Development of School–College/University Partnerships

Most educators and counselors are familiar with K–12 college/university partnerships. The collaboration is usually based on the needs and practices of both the school and the college/university. The rationale for professional development schools (PDS–school–college collaborations) developed around 2002. The concept of PDSs emerged from this call for change in the following requirements: nurturing of a learning community, a true collaboration both within and between partners, and accountability for teaching and learning. Such partnerships may provide valuable benefits for all concerned by addressing the needs and practices of both the school and the college/university. In M. Levine and R. Trachtman (Eds.) *Making Professional Development Schools Work: Politics, Practice, and Policy,*[6] they state that developing these partnerships, most importantly PDSs, must "break ground, build a foundation, and keep house" to renew their education in the 21st century. Furthermore, Fitzpatrick, C., and Mucciardi, M. (2003): *Breaking Ground in Technology Development: A School/College Partnership Focusing on the Infusion of Standards—Based Technologies into Practice,*[7]confirm that such collaborations have much to offer all involved—from the K–12 schools to the colleges/universities.

In regard to the rationale for partners noted above, there is still a compelling argument for reexamining prior partnerships, who should be in them, and where colleges, universities, and schools should proceed. Quite simply, school counselors, school psychologists, and perhaps in some cases, social workers should be active participants in the partnerships. In addition, the focus should include the development of our middle school students and their opportunity to meet with the important partnerships. The time has again come for school counselors to be proactive in developing partnerships related to their work with students and in taking leadership positions. Middle school counselors and teachers need to be proactive in developing these partnerships with colleges and universities to enhance the overall experience for middle school students that will certainly benefit their ongoing development.

An Urban School Partnership: IN-Tech Academy (MS/HS 368) and Manhattan University

Manhattan University is a private coeducational university with a graduate division situated in Riverdale, a section of the Bronx borough of New York City. The School of Education has always addressed the needs of the community for teachers and counselors as well as those of urban educators who looked for advanced degrees and in-service training. There is a commitment on the part of the Manhattan University faculty to encourage its students to consider a career (e.g., teacher, counselor, principal, guidance counselor) in an urban school setting.

IN-Tech Academy is a 6–12 NYC Public School. Since its inception, IN-Tech Academy strives to incorporate various technologies in the classroom to enhance student engagement and learning. The partnership with Manhattan University began at the earliest stages in 2000 with Manhattan University faculty involvement with faculty from IN-Tech. Dr. Corine Fitzpatrick, a professor at Manhattan University at that time, was able to get a federal grant to help develop IN-Tech in 2000. Furthermore, Kathleen Costantini developed the first advisory system at IN-Tech Academy and worked closely with the educators at IN-Tech in the ongoing development of the advisory system. Moreover, the middle school students have the opportunity in advising sessions to acquire the requisite study skills and habits, and to learn about career readiness, and to explore the world of colleges and universities. In addition, Dr. Fitzpatrick has worked with professors in the School of Engineering and the Computer Science Department. Undergraduate students from both engineering and computer science have in the past worked with teachers and students at IN-Tech Academy. The partnership between Manhattan University and IN-Tech continues at this time. The middle school programs have been very successful in achieving academic success.

College/University and Foundation Collaborations

College and university collaborations with foundations benefit the critical development of middle school children (Fitzpatrick, Mucciardi, & Sierra, 2005).[8] Once again, college and university counseling faculty are not experts, and middle school counselors and teachers are not learners. Rather, each member of the learning community brings certain strengths to the relationship. College and university faculty should participate on committees of middle schools as well as high schools and elementary schools. School counselors likewise should advise college and university faculty on best practices. For example, Dr. Fitzpatrick has worked with professors in the School of Engineering and the Computer Science Department. Undergraduate students from both engineering and computer science have in the past worked with teachers and students at IN-Tech Academy, a public technology based middle–high school. Counselors and teachers should also consider the following foundations: the Barr Foundation has noted that while their main focus is for high school students, they believe that the journey toward college and career readiness begins well before the high school years.[9] The Annenberg Institute for School Reform (Brown University) and the Center for Youth and Communities (Stanford University) received funding in 2013 from the Gates Foundation[10] to develop sites and models for College Readiness Indicators. College/university and K–12 school system collaborations with foundations can be highly beneficial to middle school student growth and development.

How Funds Help Middle School Students

While this particular fund is located in the New York City metropolitan area, many funds exist throughout the country. New York Edge, a non-profit fund, serves middle school students in afterschool and weekend programs in New York City and Long Island. Dr. Fitzpatrick has been a member of the Board for many years. We now include some of their work.

New York Edge[11]

When the last school bell rings for the day, Rachael Gazdick (Ed.M."99") gets to work. She is the CEO of New York Edge, the largest provider of afterschool and summer camp programming in New York City and Long Island, which offers daily activities for more than 25,000 students across NYC's five boroughs. Projects range from arts to sports and wellness to college and career readiness. Gazdick calls the 30-year old non-profit a "hidden gem" for the metropolitan area. Since taking over in 2019, she has worked to elevate what afterschool time can look like by creating pathways for middle schoolers to imagine the kinds of adults they might become. New York Edge and its projects provide the students with opportunities to build the confidence to make their dreams a reality.

Middle school students work directly with award-winning children's book author Jesse Byrd to learn about the writing process before writing their own stories. They are also paired with professional illustrators with different styles who illustrate the student books and bring them to life. New York Edge and its projects provide the students with opportunities, and most importantly, the same confidence to make their dreams a reality. Harvard Ed. Magazine has featured New York Edge Student Book Publishing.

Gazdick has stated that: "There is one thing about learning, there's another about helping NYC students start the school year strong. Exploring through literacy, STEM, and supplementary academic activities, New York Edge helps students make the most of their academic potential."

We thank Rachael Gazdick for all her work with NYC and Long Island middle schoolers who are able to engage in the activities offered at New York Edge.[12]

Finally, we present a planner, "Your First Step towards Reaching Your Goals" from ***KIDS2College*** provided by The Sallie Mae Fund. We are also including this *First Step towards Reaching Your Goals* here for counselor and teacher consideration of this fund as an example of an excellent source of information and resources as this document coincides with much of the work of the advisory/group counseling program.

Advisors and counselors should be ready to explain the following steps.

"KIDS 2College: Middle School: Your First Step Towards Reaching Your Goals"[13]

It might sound like everything you need to do to get ready for college happens in high school. That is not the case. *KIDS 2College* has provided a list of things that middle schoolers can do right now as a middle school student to get on track for college.

Sixth Grade

1. Make sure you know your learning style—do you learn better seeing, hearing, or doing?
2. Develop good study habits based on your learning style.
3. Be sure to attend school regularly, take it seriously, and get good grades.
4. Do your homework—homework is practice, and this is how you improve.
5. Keep talking to your parents about what you are doing in school each day.

Seventh Grade

1. Participate in school activities such as music or sports.
2. Be active in clubs if you have the opportunity.
3. Keep doing your homework!
4. Talk to your counselor about what you need to do to be ready for algebra in eighth grade.
5. Continue looking at career options that interest you.
6. Review career information (pamphlets, brochures, websites) to see what general types of jobs might interest you.
7. Talk to your family and friends about career possibilities. If an adult that you know works in a career that interests you, see if you can spend a day job shadowing—going to work with them for a day to see firsthand what they do in their career.

Eighth Grade

1. Continue to develop very good study habits and organizational skills.
2. Take Algebra if it is offered at your school.
3. Keep up with your homework!
4. Keep working hard and getting good grades. It is important that you do well in ALL of your subjects, not just your favorites, so that you are ready to go for high school.

5. Talk to your teacher and guidance counselor about how you are doing. Ask them for guidance in selecting you ninth-grade courses.
6. Talk to your parents about how your high school classes, college plans, and career goals all fit together.
7. Have your parents learn about financial aid opportunities that can help them pay for your college education.

This planner is an excellent guide that all middle school students should have. While middle school may be years away from college, this planner provides practical advice that can help all middle schoolers!

Conclusion

In conclusion, all participants benefit from collaborations. Colleges and universities have the opportunity to work directly with middle school educators, with both sides of the collaboration learning from each other. Hopefully, these partnerships will result in enhancing academic achievement, career and college readiness, and social/emotional development of middle school students. Furthermore, foundations such as the Barr and the Gates Foundations provide essential assistance to middle school students. Finally, funds also play a key role in the growth of adolescents. New York Edge and the Sally Mae Fund provide extensive assistance to middle school student growth and development. Such partnerships and collaborations benefit all involved.

Notes

1 *ASCA School Counselor Professional Standards & Competencies*. American School Counselor Association, p. 1.
2 *ASCA School Counselor Professional Standards & Competencies*. American School Counselor Association, p. 1.
3 *ASCA School Counselor Professional Standards & Competencies*. American School Counselor Association, p. 1.

4 Pappas, Stephanie. "Partnering with Schools to Strengthen Education Research." *American Psychological Association*, May 2025. https://www.apa.org/research-practice/conduct-research/partnering-schools-strengthen-research
5 "Research and Practice." *American Psychological Association*. https://www.apa.org/research-practice
6 Levine, M. and R. Trachtman (Eds.). *Making Professional Development Schools Work: Politics, Practice, and Policy*. New York Teachers College Press, 1997.
7 Fitzpatrick, C., and M. Mucciardi. *Breaking Ground in Technology Development: A School/College Partnership Focusing on the Infusion of Standards—Based Technologies into Practice*. Eric Publication, 2003.
8 Fitzpatrick, C., and M. Mucciardi. *Breaking Ground in Technology Development: A School/College Partnership Focusing on the Infusion of Standards—Based Technologies into Practice*. Eric Publication, 2003.
9 "Education." *Barr Foundation*. https://www.barrfoundation.org/education/
10 Gates Foundation. https://www.gatesfoundation.org/
11 New York Edge. https://newyorkedge.org/
12 New York Edge. https://newyorkedge.org/
13 "Kids2College The Sallie Mae Fund Teacher's Edition," p. 76. *Scribd*. https://www.scribd.com/document/705361523/k2c-teacher-intro-COVID-19

14

Working with the Families of Middle School Children

As young adolescents enter middle school, they encounter multiple challenges psychologically, socially, and emotionally, as well as the heightened levels of critical thinking that academic achievement demands. How can we as teachers, counselors, and administrators work with parents to assist our students for optimal growth in this critical time in their young lives? One crucial practice is to ensure that avenues of communication are always open in this all-important stage of adolescent development.

In the first section of this book, the focus is on the new academic challenges, psychological development and social-emotional well-being of middle schoolers. While we normally think of Parent Meetings as pertaining to academic issues, why not consider group meetings for parents as a way of presenting valuable insight into their child's psychosocial development? At minimum, parents should be aware of whom to contact at the middle school if they believe that their child is experiencing any serious psychological, social-emotional, or academic issues. Always provide open forms of communication for parents.

DOI: 10.4324/9781003597292-17

Why Communicating with Middle School Parents Matters in Helping Students Achieve Greater Academic Success

As students move into middle school, parents too may find themselves entering a new world of academic challenges. Particularly as learning management systems have become a prevalent feature of middle school, parents may feel that they too may be in need of some guidance as they oversee homework completion and test preparation, not to mention the expanding world of their children's activities and social interactions.

For many middle schoolers, their world has now expanded as they meet new people, both students and teachers, change classrooms for subjects, use lockers, eat lunch at different times on various days, experience new afterschool clubs and sports; even how they travel to school may be different! With these changes come parental concern about their child's adaptability to adjust to the multifaceted environment of middle school. Even for seasoned parents who have experienced this phenomenon with an older child, the world of Learning Management Systems has profoundly changed the landscape of the middle school experience.

Parental involvement is essential to the success of middle school children. As students progress in developing goals, enhancing study skills and time management strengths, and exploring careers and colleges, parents need to understand how these activities are positively impacting their child's life. Communicating with parents is an integral part of the Advisory/Group Counseling Program—group meetings–individual meetings—the key word is meetings! Administrators, teachers and counselors should determine a methodology of communicating with parents that works best to facilitate this communication in their individual schools.

The Importance of Middle School Psychosocial Development for Parental Understanding

Middle schoolers who are entering adolescence develop further into identity formation, including a sense of their awareness about their interests, strengths, weaknesses, and beliefs. Such a focus on identity development was first noted by Erik Erikson who is known for his psychosocial theory of development. Erikson's theory recognizes the effect of external factors, parents, and society on personality and development that spans over eight stages from childhood to adulthood. In Stage 4, which spans from six to twelve years, children are now at the social stage. The child's parents are still integral, but the world is larger in relationships with school and community. At this point, self-esteem and competency issues may develop. Furthermore, if children experience support from parents and the school community, they are more able to accomplish goals.[1] Clearly, Erikson's theory has influenced further research into psychosocial development. Middle schoolers have learned to depend on various systems to foster their social, emotional, personal, and developmental needs. These systems include their new peers, their teachers, their families, and their community.

Also, as the Bowen family systems theorizes, adolescents may begin to differentiate from their families, which may strain the family system. At this time, they begin to form identity more specifically and begin to include their sense of awareness about their interests, strengths, weaknesses, and, in some way, their beliefs.[2] Finally, James Marcia presented four possible identity statuses for developing middle school adolescents that comprise the relationship between exploring their future ideas and developing their identity: identity diffusion, identity foreclosure, identity moratorium, and identity achievement.[3] We believe that understanding these developments would be very helpful for discussion with parents. Again, as suggested above, why not host a Parent Meeting in which a school psychologist speaks to parents about the psychological development that is a normal part of their child's growth?

Furthermore, Social–Emotional Learning (SEL) and Mental Health Challenges provide a wide spectrum of professionals who need to be engaged in middle school development. For SEL to work, adolescents must engage with each other. Bandura's self–efficacy theory stresses the importance of a group framework for social learning which offers a positive interaction in psychosocial development overall. Parents need to understand the power of group social learning. Parents may want to understand examples of SEL which focuses on how children interpret social cues and social challenges. A discussion of Bandura's framework for social learning includes self- efficacy as playing a pivotal role in adolescent development.[4] Also from Chapter 3 is a discussion/review of SEL and the Mental Health Challenges that may arise.

Finally, as discussed in Chapter 3, parents should be aware of an article in the *New York Times, Surgeon General's New Mission: Adolescent Mental Health.* Although this interview occurred in 2023, the Surgeon General's letter was issued in December of 2021 as a rare warning: "Mental health challenges were leading to "devasting effects among young people."" Why should that be a consideration for discussion with parents? Parts of that "Mission" might be well worthwhile to present to parents about such challenges. Murthy believes that young people, especially in "early adolescence," are in a "very sensitive phase of development." Murthy adds that while adolescents are spending so much time online, that time is taking away from their sleep, from exercise, from in-person interaction with people, from schoolwork and from other activities that might bring them joy. Dr. Murthy ended his discussion accordingly: "Many of them feel exhausted…Are we pushing our kids to pursue what's really going to bring them happiness and their fulfillment?" It is essential to keep in mind that the Surgeon General's report is not just based on statistics. Murthy has come to these conclusions after a lengthy listening tour speaking with students, health care workers and community groups across the country. His service is remarkable in defining what is not clearly a hazardous, risky behavior such as smoking, but rather a mental health crisis that is afflicting our nation.[5]

Furthermore, the American Mental Health Counselors Association might be very helpful to parents who need to find a clinical counselor or psychologist. An added note: Dr. Fitzpatrick, as a licensed Clinical Psychologist and as a licensed Mental Health Counselor, has been able to advocate for the recognition of having Mental Health Counselors working in school settings. Her review of the "Recommendations for the Role and Responsibilities of School—Based Mental Health Counselors" article in the Journal of School–Based Counseling Policy and Evaluation (Christian, D.D., & Brown, C. L. (2018)[6] has definitely noted the authors' recommendations from 2018. Having licensed Mental Health professionals service in middle and high schools is truly needed and should be part of any set of services delivered to a school setting.

While the sample agendas presented below follow the step-by-step shared agenda of the advisory/group counseling programs, counselors, teachers, and administrators should decide on what form of communication may best address the needs of the school community in regard to the psychological development and social and emotional well-being of students. Group meetings for parents are a way of presenting valuable insight into their child's development. In these meetings, the school psychologist (if one is on staff), the head of the school counseling department, or perhaps a guest psychologist may address these developmental issues with parents. A sample agenda should include a discussion of adolescent development as presented above.

What Follows Are Sample Agendas for Group Meetings with Parents

Why not conduct a Parents Assembly to introduce parents to life in middle school? Whether parents have experienced middle school already or are new to this key educational step in their children's lives, meetings are fundamental to ensuring that parents understand both the demands and the growth opportunities that are part of the middle school experience. Particularly for those schools using a Learning Management System such as Google Classroom or Schoology, parents need an introduction to how this system impacts both parents and students. Principals may also provide language interpreters as needed.

The sample agendas presented follow a step-by-step process designed to help middle school parents understand what is happening in the academic life of their child. Included is the presentation of the Advisory/Group Counseling Program, which may be new to many parents. While the following agendas provide a framework for group parent meetings, periodic advisor/counselor individual meetings with parents and students are an excellent way to include parents on what is happening as students progress through the middle school years. Administrators, teachers and counselors should decide upon a common framework for such meetings, that is, once/quarter or semester.

The opening agendas for a two-part initial parent meeting may be adapted for a one meeting approach. However, the introduction of a Learning Management System (LMS) may significantly lengthen the meeting.

SAMPLE AGENDA OF THE INITIAL PARENTS MEETING

(In early September, for incoming middle school students)
Aim: To introduce parents to the middle school

Agenda of Meeting 1

Part I

Principal welcomes parents in an auditorium or gym setting. In this presentation, the principal should briefly describe how middle school differs from elementary school:

- For some students, taking a bus for the first time.
- For most students, interacting in a much larger student body.
- Meeting new teachers.
- Students change classes.
- Use of lockers.
- More activities and sports.

Principal then introduces Assistant Principals who briefly describe their role in the middle school student's life followed by the introduction of the Head of the School Counseling

Department who briefly describes the role of counselors in the life of sixth graders.

Part II

Parents visit lockers and homerooms.

- In some schools, lockers may be several floors away from a student's homeroom. Students may need to become more organized in planning for their morning classes.
- After seeing the locker area, parents then go to their child's homeroom where they will meet the homeroom teacher who describes homeroom activities. Including: taking attendance, school announcements, etc.

Part III

Parents then return to the auditorium/gym setting for the principal's concluding remarks involving student adaptation to life in middle school. The principal concludes by explaining the preferred method of parents communicating with staff (phone/email).

This initial Parents' Meeting should be followed by a second meeting that has a two-fold purpose: to introduce the LMS and to provide parents with an opportunity to meet the subject teachers.

SAMPLE AGENDA OF THE SECOND PARENTS MEETING

(To follow the Part 1 Meeting Within Two Weeks)

Aim: This meeting has a twofold aim: to introduce parents to both the LMS and to their children's subject teachers.

Part I

Principal welcomes parents in an auditorium/gym setting. In this meeting, parents learn essential features of the LMS. (Systems usually provide a video presentation explaining the purpose and functioning of the LMS.)

Key Features in Google Classroom and Schoology

- Access to learning materials: students can access various learning resources, including lessons, videos, and documents, all in one place.
- Assignment Submission and Tracking: students can submit assignments, track their progress, and receive feedback from teachers.
- Gradebook and Progress Reports: students can view their grades and progress reports, allowing them to monitor their performance.
- Communication and Collaboration: facilitates communication and collaboration through discussions, forums, and messaging, fostering a sense of community.
- To-Do List: students can see a list of upcoming assignments and deadlines.
- Aggregated calendar: students can see their assignments and events in one place.

(More detailed descriptions of Google Classroom and Schoology are provided in Chapter 6.)

*Recommendation regarding the use of traditional planners. Students may find it helpful to use a hard copy planner to aid them in developing time management skills in this new environment.

Part II

Parents meet the subject teachers. Following a copy of their child's schedule for a given day, parents go to each classroom. Here teachers will introduce themselves and give a brief description of the subject matter. (Each teacher session should not exceed 10–15 minutes.)

Part III

Parents return to the auditorium/gym setting for the principal's concluding remarks involving middle school life in the new age of the LMS.

For schools not using a LMS, the Part I of the meeting should cover:

- Using a traditional planner for assignments.
- Timely submission of assignments.
- Interim progress grade reports.
- Communicating with the school administration.
- Communicating with faculty and counselors.

Parts II and III Remain the Same.

As students settle into life in middle school, hopefully they will meet in advisory or group counseling sessions. In these meetings, students move through the steps of understanding executive functioning and how it relates to accomplishing goals and enhancing study skills and time management. Parents of sixth graders need to be familiar with the advisory/group counseling sessions and how these sessions positively impact their children.

SAMPLE AGENDA OF PARENTS MEETING

(should be held in early to mid October)

Aim: This meeting has a twofold aim: To introduce parents to both the Advisory/Group Counseling Program and to their children's advisor/counselor.

Part I

Principal and the Head of School Counseling Department welcome parents in an auditorium/gym setting. In this meeting, the Advisory/Group Counseling Program is presented to parents.

- Defines the advisory/group counseling program.
- Who are the advisors/counselors and frequency of sessions.
- Initial Objectives: executive functioning, establishing goals. Enhanced study skills, time management, and avoidance of distractions.

- Dangers of doomscrolling and excessive social media viewing.
- Students must do their own homework—not AI or ChatGPT.
- Importance of a quiet homework/study space free of distractions.

Part II

Parents attend their child's advisory/counseling session. In this session, the advisor/counselor should explain the steps of understanding executive functioning and how it relates to accomplishing goals and enhancing study skills and time management. Parents should also become familiar with the exercises and charts that their children have been using to develop enhanced study skills and time management.

If the school uses a LMS, include a discussion of the To Do List and Calendar. In addition to these features, students may choose to use a hard copy planner as well as the LMS features.

For schools not using a LMS, include a discussion of the necessity of using a hard copy planner for both daily and long-term assignments. *

Part III

Parents return to auditorium/gym for concluding remarks involving the Advisory/Group Counseling Program.

*Advisors/counselors may want to select the best planner for students and distribute that planner when school opens or when students receive textbooks.

As students progress, activities in the Advisory/Group Counseling sessions will advance to career and college exploration. Once again, parents need to understand how their middle school children are engaging in this process.

Another Suggestion: Why not create a homework video?

Particularly for the first-time parents of students entering middle school, understanding the importance of creating a quiet, orderly homework space for their child is essential. Posting a brief video on the school website narrated by an administrator, counselor, or teacher may be beneficial to parents and students alike. If not a video, posting guidelines for a good study space is another option.

Suggested script:

Advisor/counselor introduces the Homework Video and Why a Quiet Homework Space is Necessary.

- Students in middle school are engaging in a higher level of critical thinking.
- Being prepared for class is essential.
- Students need a quiet space in their home where they can do homework and study.
- Your children are learning the difference between homework and studying in advisory/group/counseling sessions. Both homework and studying need a quiet space!
- Set aside an area that is a study space for your child.
- If students have their own bedrooms, establish a desk or table space for them.
- If they share a room with a sibling that is fine; just make sure that siblings understand the concept of a quiet homework space.
- If necessary, establish a study space in a common room: in a dining room, at the kitchen table, wherever students can focus.
- Don't allow distractions: turn off cell phones! No doomscrolling or visiting social media sites!
- Build in breaks!

Advisor/counselor concludes by stressing that middle schoolers are engaging in a higher level of critical thinking and how the advisory/group counseling program is helping them in this developmental process. A final concluding note is to advise parents to communicate with advisors/counselors about concerns that they may have regarding their child's progress. (Suggest the best method of communication: LMS, email, or phone.)

Next Step……

As students progress, activities in the Advisory/Group Counseling sessions will advance to career and college exploration. Once again, parents need to understand how their middle school children are engaging in this process.

SAMPLE AGENDA OF PARENTS MEETING

(This agenda may be used for sixth –eighth grade students as appropriate. For example, some sixth graders may have started career exploration in the second semester of sixth grade. Entering seventh grade parents may benefit from an update on career exploration.)

Aim: To introduce parents of sixth–eighth grade students about the Why behind career exploration in middle school and the How students are pursuing this activity.

Part I

Principal and Head of School Counseling welcome parents in an auditorium/gym setting. In this meeting, they explain the motivational factor involved in exposing middle school students to career paths at a young age. (See Chapter 8.)

Part II

Parents attend their child's advisory/group counseling session where they are introduced to the career exploration method used (for purposes of this sample agenda, the ACT which allows students as young as eleven to create an ACT account will be the working example).

- Students have created an ACT account.
- Explanation of the Inventories: Interests, Abilities, and Values.
- Explanation of the Career Map.
- Research through the Career Map.
- Sharing of Exercises that students are/will be using.

Part III

Parents return to the auditorium/gym for Principal's concluding remarks about the importance of early exposure to career exploration.

Why not think about conducting a **Career Fair** for middle school students? This meeting would be an opportune time to announce a Career Fair. (See Chapter 8 for more details.)

As students complete their initial examination of possible careers, the next step is college exploration. Again, for purposes of this sample agenda, the ACT will be the website for college exploration. (See Chapter 9)

SAMPLE AGENDA OF PARENTS MEETING

(This agenda is appropriate for the parents of seventh–eighth grade students who are moving into the college exploration step of the Advisory/Group Counseling Program.)

Aim: To introduce parents of seventh–eighth grade students about the WHY behind college exploration and the HOW students are pursuing this activity.

Part I

Principal and Head of the School Counseling welcome parents in an auditorium/gym setting. In this meeting, the Principal and Head of School Counseling Department explain the motivational factor involved in exposing middle school students to college exploration at a young age. (See Chapter 9.)

Part II

Parents attend their child's advisory/group counseling session where they are introduced to the college search method used (for purposes of this sample agenda, the ACT which allows students as young as eleven to create an ACT account will be the working example).

- Introduction to the Major Map—what majors are needed for specific careers.

- Learning about liberal arts colleges.
- Understanding what a university is.
- What degree is needed for a specific career.
- Visiting colleges and universities on websites.
- Brief introduction to financial aid options.
- Sharing exercises that students are/will be using.

Part III

Parents return to auditorium/gym for Principal's concluding remarks about the importance of early exposure to college exploration. (See Chapter 9 for more information.)

As middle school students move into the second semester of seventh grade/first semester of eighth grade, the question of what high school to attend looms on the proverbial horizon. For many students, attending their zoned high school is the next step in their education. However, for others, applying to a specialized public high school, an independent school—both private and Catholic—may be choices that students and parents want to pursue. Particularly in large urban areas, multiple possibilities exist for middle schoolers to explore. Parents and students need to investigate options available to them. How can the teachers, counselors and administration help in this pursuit? In this **Sample Agenda of a Parents Meeting** that leads into a high school style fair, families may visit representatives at booths set up in the gymnasium or in classrooms to learn more about high school options. This agenda may be adapted to an information agenda that does not include the high school fair portion of the meeting. What is of key importance is that both seventh and eighth grade parents and their children attend this meeting.

SAMPLE AGENDA FOR PARENTS MEETING ABOUT HIGH SCHOOL POSSIBILITIES

(This agenda is appropriate for seventh–eighth grade families.)

Aim: To introduce families to high school options (This meeting also includes a high school fair.)

Part I

Principal welcomes families in the auditorium and explains the "basics" of high school admission.

- Zoned schools.
- Specialized high schools (if applicable to school district)

Principal may ask a Board of Education/District speaker to make a brief presentation.

- Entrance exams for specialized high schools and test preparation information.
- Entrance exams for Independent/Catholic schools and test preparation information.

Part II

Families visit high school booths set up in the gymnasium where representatives of high schools provide information about their schools. An alternative method would be to assign classrooms to high school representatives to make presentations.

Part III

Families return to the auditorium for the Principal's concluding remarks about the high school admission process. (See Chapter 10 for more information.)

Ongoing Communication with Parents

While the agendas presented cover the key steps that advisors/counselors are pursuing in their sessions with middle school students, it is important to keep communication open between advisors/counselors and parents on an individual basis. In schools using a LMS, parents may contact the advisor/counselor through a portal just as they would reach out to individual teachers. Email and phone contact are options for schools not using a LMS. The essential factor: ongoing communication with parents is an integral part of the Advisory/Group Counseling System.

Furthermore, as advisors/counselors continue their work with eighth graders about high school options, it is essential to include parents in this ongoing process. In addition to the Parent/Student Meeting described above, we also suggest the following ways to include parents in this all-important process. (See Chapter 10 for specific information.)

SAMPLE AGENDA FOR PARENTS MEETING ABOUT HIGH SCHOOL POSSIBILITIES FOR EIGHTH GRADE STUDENTS

(This agenda is appropriate for first semester eighth grade families. This meeting should be held in early September.)

Aim: To make sure that families understand high school options and what is required for high school admission. Most importantly, this meeting is an ideal time to ascertain that parents and students are on the right course for high school.

Part I

Principal welcomes families in the auditorium and reminds parents of the basics of high school admission. (Again, seek assistance from the District Superintendent/Board of Education in presenting public high school possibilities. Also, representatives from private schools may present information about the application process.) Most importantly, this meeting is an ideal time to make sure that all students are on the right course for high school

- Zoned schools.
- Specialized high schools (if applicable to school district).
- Entrance exams for specialized high schools and information about test preparation.
- Entrance exams for Independent/Catholic schools and information about test preparation.
- Possible visits to high schools to which students are applying: Open Houses, day visits, etc.

Allow sufficient time for a Q and A session.

Part II

Principal presents a copy of **Exercise 14.1** and explains that students will bring this exercise home in early to mid-October for students and parents to complete. Exercise 14.1 will be presented to students in the Advisory/Group Counseling Program as they explore high school options. Exercise 14.1 may be helpful in ascertaining that parents and students alike understand high school options.

EXERCISE 14.1 My High School Options (to be completed by student and signed by parent)

Name:

Part I
List High Schools in order of preference:

Is special testing required?

What is the test?

How are you preparing for the test?

Part II
What are your other high school options if you do not gain entrance to your first choice option?

Is testing required?

If so, how are you preparing for the test?

Part III
Is there a zoned or neighborhood school that you would like to attend?

Signature of parent:

Part III

Principal concludes the meeting with a reminder that parents should not hesitate to contact advisors/counselor/administrators if they have any questions.

Advisors/counselors may also conduct individual meetings with students and parents in both seventh and eighth grades. As in the Parents Meeting format, students and parents should be informed about high school options, any testing requirements, and suggestions for test preparation. Like the Parents Meeting format, these individual meetings are a great way to ensure that all eighth graders are finding their best high school possibility.

Whether students attend their zoned/neighborhood school, a specialized high school, or a private/Catholic high school, what is important is that they are involved in this key process that will lead to the next step in their education.

A Final Note....

From sixth grade through eighth grade, parents and students alike will benefit from the ongoing interactions of meetings with the educators in their child's middle school. It is important that parents understand the work of the Advisory/Group Counseling sessions as their children progress into becoming proactive students who are also venturing into career and college exploration. Especially as parents and students investigate high school possibilities, meetings are a natural extension of this ongoing process.

*In Chapter 15, we present an agenda for helping parents and students alike as they transition to high school.

Notes

1 Cherry, Kendra. "Erikson's Stages of Development." *Verywell Mind,* November 2025. https://www.verywellmind.com/erik-eriksons-stages-of-psychosocial-development-2795740

2 Strauss Cohen, Illene. "Understanding Bowen Family Systems Theory." *Psychology Today,* November 2023. https://www.psychologytoday.com/us/blog/your-emotional-meter/202311/understanding-bowen-family-systems-theory

3 Marcia, James, et al. *Ego Identity: A Handbook for Psychosocial Research,* 2nd edition. New York: Springer-Verlag, 1993.
4 Bandura, Albert. *Social Foundation of Thought and Action: A Social Cognitive Theory.* Prentice Hall, Inc, 1986.
5 Richtel, Matt. "The Surgeon General's New Mission: Adolescent Mental Health." *New York Times,* March 21, 2023. https://www.nytimes.com/2023/03/21/health/surgeon-general-adolescents-mental-health.html
6 Christian, David D. and Cian L. Brown. "Recommendations for the Role and Responsibilities of School-Based Mental Health Counselors." School-Based Mental Health Counselors. *Journal of School-Based Counseling Policy and Evaluation,* 2018, Vol. 1, pp. 26–39.

15

The Move to High School

Making Transitioning to 9th Grade Successful: Second Semester 8th Grade Advisory/Group Counseling Sessions

As eighth grade students enter the second semester, they are already in the early stages of transitioning from middle school to high school. For those students who have applied to various high schools and are awaiting decisions, high school has been a constant factor in their young minds. As we discussed in Chapter 10, many eighth grade students may have multiple possibilities in applying to specific high schools that may emphasize an area of study that they may wish to pursue. One of the benefits of the early career exploration as presented in Chapter 8, is that students have had an early experience of taking the ACT's Inventories—Interests, Abilities, and Values—which may influence their choices in applying to high schools. For example, in New York City, the Bronx High School of Science may attract students who know that the many areas of science are what they want to pursue while others may apply to the High School of American Studies at Lehman College to pursue their interests in American history, literature, and culture. However, many eighth graders are in districts where high school choices are more limited as all middle school students move onto one high school. No matter where eighth graders attend high school, at a nearby larger school building or traveling on the subway into what to

DOI: 10.4324/9781003597292-18

them is a new borough of New York City, they will all experience a period of transition to their new environment. Just as young Tricia felt excited and yet a bit anxious as she traveled on the school bus taking her to the new middle school, entering ninth graders will experience many challenges as they transition to high school. Students who leave the eighth grade and enter their first year of high school are at a critical point in their young lives. Helping these students move to this new stage of their K–12 journeys is so important in their successful transition to high school.

Adapting to their new high school environment is a significant factor in the lives of young ninth graders. While some middle schools provide little transitioning assistance, the same may be true for some high schools. Too often, ninth graders must adjust to the many new challenges of high school with minimal school guidance. In a study, Kerr (2001) found that at large comprehensive high schools in particular, little support was available for new ninth graders. More recently in her dissertation, Simisola K. Aromolaran focused on the challenges that many graduating eighth students experience as they enter high school. The results of this study indicated that students felt overwhelmed, especially during the first few weeks of transitioning to ninth grade. As noted by Aromolaran, this transition period occurs at a time when many adolescents experience various cognitive, psychological, and physiological changes. Although the study involved a relatively small sample, five student participants and five parent participants, and occurred during the COVID-19 pandemic, the evidence demonstrates positive outcomes. The entering students quickly overcame challenges as they utilized available support provided by school educators. Clearly, these findings demonstrate that parents and school educators played a role in ensuring that the transition was smooth and manageable.[1]

High school counselors and teachers may also assist students in making a successful transition to the academic demands of secondary school. As presented in *Easing the Transition to High School: An Investigation of Reform Practices to Promote Ninth Grade Success*[2] the use of various inventories such as preferred learning styles may deepen their awareness of the learning process and their own growth as they transition to ninth grade. As noted in

Chapter 6, EDUCATIONPLANNER.org provides a 10-question survey that helps students to understand themselves as students. This site also provides a 20-question survey that helps students to understand their preferred learning style.[3] Students may naturally learn to adapt their preferred learning style to specific subjects; however, just knowing what a learning style is may help them adjust to the various subjects that they will take in high school. School counselors should make this part of the transitioning process to high school.

While high schools need to address transitioning, middle school administrators, teachers, and counselors must prepare students and families during the second semester of eighth grade, prior to the move to high school. Before we provide transitional activities that can be used to help students in this all-important process, we refer to Chapters 2 and 3 to provide a short developmental overview, focusing on Psychosocial Development and Cognitive Development for eighth grade students.

Psychosocial and Cognitive Development

As eighth graders enter high school, they may struggle to identify who they want to be and consequently may become confused. As these adolescents endeavor to identify who they want to be now and in the future, they may question their beliefs, values, and their possible future careers. Additionally, eighth graders often become highly influenced by their peers and want to be part of a particular group. The family influence is still important, but peers are very influential.[4] At the same time, students in eighth grade are more sophisticated in their thinking.

> Additionally, abstract thinking gives adolescents new skills to use, including thinking logically, engaging in deductive and inductive reasoning, generating hypotheses and problem solving. These skills are essential to career and college planning, as doing so requires the

> ability to understand one's options, research possibilities, comprehend one's own talents and abilities, and project into the future.[5]

An essential part of the advisory/group counseling sessions has been an initial career and college exploration, helping middle school students to imagine and explore their possible futures. For ninth graders who have experienced the step-by-step approach of the advisory/group counseling program, their transition to high school may occur quite naturally as they have engaged in a developmental program that is based on students acquiring the foundational basis of strong study habits and skills as well as an initial exploration of careers and colleges.

Clearly, eighth grade students as they transition into high school are truly entering into another phase of psychosocial and cognitive development. We as teachers, counselors, and administrators can assist them in this all-important process. The ASCA Mindsets and Behaviors for Sixth- and Seventh-Grade Curriculum Planning Focus and the ASCA Mindsets and Behaviors for Eighth-Grade Curriculum Planning Focus as noted by Curry and Milsom may be helpful in the transition process.[6] As we proceed into the advisory/group counseling sessions, we review, in particular, Category Two: Behavior Standards as a way of assuring eighth graders that what they have experienced in the advisory/group counseling sessions has been a solid preparation for high school and beyond.

ASCA Mindsets and Behaviors for Eighth-Grade Curriculum Planning Focus

In particular, the ASCA Mindsets Standards describe the self-confidence and beliefs in themselves that an eighth grader should have in their abilities to achieve high-quality results and outcomes. The Behavior Standards fall into three categories:

Learning Strategies include the abilities to:

- Use time management, as well as organizational and study skills.
- Identify long- and short-term academic, career, and socioemotional goals.

Self-Management Skills include:

- Demonstrate self-discipline and self-control.
- Demonstrate the ability to manage transitions and the ability to adapt to changing situations and responsibilities.

Social Skills include:

- Demonstrate ethical decision-making and social responsibility.
- Demonstrate social maturity and behaviors appropriate to the situation and environment.

Eighth grade students who have truly attained the Mindsets and Behaviors Standards are ready to engage in the higher demands of critical thinking that high school will demand. In the advisor/group counseling program, we have worked systematically to help middle school students establish a solid foundational basis that is essential in reaching the ASCA Standards, and indeed the College and Career Readiness Standards. At the same time, there has been a clear focus on goal establishment and accomplishment. As students have developed their self-management skills, they have learned to adjust to changing circumstances and responsibilities. Working in group settings and participation in extracurricular activities and sports has further advanced their social skills. At this young age, they are capable of making ethical decisions and understanding consequences. Following the rules of the middle school encourages solid decision-making as students may encounter opportunities to cheat in academic assignments. Students understand the consequences for such behavior as they have learned to do their own work whether they are working individually or in a group setting. Finally, these skills also demonstrate that their social maturity and behavior are appropriate in given situations. As in any given situation, there may be individuals who may need more assistance in achieving the ASCA Standards. What Curry and Milsom present in their discussion of the Mindsets and Behaviors Standards is the definition of an adolescent truly moving ahead with the necessary abilities and attributes that are essential for the K–12 journey.[7]

How We Can Assist Eighth Grade Students and Their Families in the Transition to High School

Helping both eighth grade students and their parents prepare for this transition is an important part of the advisory/group counseling program. In these final sessions, we hope to make starting high school an exciting transition, one that students who have experienced the step-by-step shared agenda of the advisory/group counseling program may find to be less challenging. What we are presenting in this chapter is a method of working with current second semester eighth grade students and their parents to prepare them for this all-important transition to high school. Parents have had previous meetings with the advisors/counselors in sixth and seventh grades discussing the topics presented below. *(See Chapter 14 for Parents Meeting Agendas.)* As we move forward in these sessions, we include parents in the transitioning process for second semester eighth graders. At the conclusion of the student section, a final Parents and Eighth Grade Students Meeting Agenda will be presented.

Assisting students and their families is the primary purpose of using the Advisory System or the Group Counseling Program presented in Part II. Now, in these final sessions, we initially present a brief review of what has been accomplished in the advisory/group counseling sessions as a means of assuring students that they are ready for high school. In these sessions, advisors/counselors remind students of the work that they have accomplished, particularly in developing strong study skills and habits and defining goals and the means to accomplish them. Furthermore, through the use of learning style and study skills inventories, we remind students that they have deepened their awareness of the learning process and their own growth in the transition process. Students have also experienced early career exploration and college search, both of which have hopefully broadened their ideas about future possibilities and what is needed to achieve their career aspirations. In essence, what we are doing in these final sessions is reminding eighth graders of what they have accomplished in the advisory/group counseling sessions and assuring them that they are indeed prepared for high school. Now, as these middle schoolers are entering a new stage of academic life, again we present a series of advisor/group counseling sessions in the voice of the advisor/counselor speaking to students in the second semester of eighth grade as they begin the transition to high school.

Part I. Helping Students in the Transition to High School (second semester eighth grade students) Advisory/Group Counseling Sessions.

Session 1

Welcome eighth graders,

By now, you may have a good idea of what high school you will be attending, and you may be experiencing mixed feelings as you are in the initial stages of the transition process to high school:

- Nostalgia about leaving what has been your home school for the last three or more years.
- Excitement about life in high school academically and socially.
- Some anxiety about the challenges ahead.
- Let's talk about how you feel, but first, complete Exercise 15.1

After students have completed Exercise 15.1, the advisor/counselor should conduct a group discussion or break students into small groups, who later share their responses in a group discussion.

EXERCISE 15.1 Transition to High School

1. How do you feel about moving on to high school?
2. Are you excited? What about?
3. Are you anxious about this transition?

A note to the advisor/counselor: keep a printed or digital record of this exercise for future reference.

The purpose of the following advisor/group counseling sessions is to remind students of the foundational work that they have done that is so essential to academic and personal success.

Session 2

A note to the advisor/counselor: What is being reviewed is the shared agenda of the advisory/group counseling sessions that also relate to the ASCA Model noted above.

What we are going to do in the following group sessions is to remind you of what you have accomplished in your middle school years that prepares you for the challenges ahead.

Let's talk about **Executive Functioning** and your own **ExF factor and the four keys to improving your ExF factor: Effective Planning, Organization, Time Management, and Avoidance of Distractions.** (See chapters 5 and 6 for details.)

Advisor/counselor should engage students in a review of these terms and question students as to how effectively they have developed the three keys.

- Do you effectively plan how you will complete homework and long-term projects?

- Are you organized on a daily basis? For long-term assignments?
- How is your time management, both daily and in the completion of long-term assignments?
- Do you avoid distractions? Cell phone off? No social media sites during study time?

After completion of discussion, the advisor/counselor should remind students that effective executive functioning is fundamental to academic and personal success.

Session 3

The purpose of the following advisor/group counseling session is to remind them of who they are as a student: what kind of student are you and what is your preferred learning style. See Chapter 6.)

Students have already participated in EDUCATIONPLANNER. org[6] *20 question survey that helps students to understand their preferred learning style and their 10 question quiz that helps them to understand who they are as students. Why not have students retake these quizzes to ascertain any changes in how they view themselves? If students have not completed these quizzes, they should do so now.*

Have students complete the following exercise.

EXERCISE 15.2 What Kind of Student are You?

What kind of student are you?	*How can you improve?*
What is my preferred learning style?	Give examples of how it helps you in school.
What do I need to be aware of when learning in a style that is not my preferred learning style?	

After students have completed this exercise, engage them in a group discussion about why understanding their preferred learning style can help to achieve personal and academic success.

Session 4

In this session, remind students of the study skills that they have acquired over the middle school years. *(Again, see Chapter 6 for details.)*

Essential Steps to Review:

- The difference between doing homework and studying.
- Why it is key to review class notes before beginning homework.
- How to take good homework notes.
- How to use bolding, indenting, and spacing to effectively structure class notes and homework.
- The Dos and Don'ts about Test Taking.

It is so important that students understand that they have a solid foundation in the fundamental skills that will promote their future personal and academic success.

Session 5 may be done in one or two sessions.

As students achieve an understanding of how the foundational skills that they have acquired in middle school will support them in high school, it is time to review what they have learned about themselves in career and college exploration.

Again, in the voice of the advisor/counselor speaking to students…

Now that you know that you have the fundamental skills and foundational basis for academic success, let's take another look at your career and college exploration. Remember, while this is a preliminary venture into these areas, you have discovered the **what** and **why** behind your career interests. You have also acquired a new vocabulary as you initiated early research into colleges and universities. *(See Chapter 8 for career exploration and Chapter 9 for college/university searches.)*

First, let's review **career** exploration on the ACT website:

- Interests, Abilities, and Values Inventories
- Career Map.
- Majors related to your interests.
- Your exploration of possible careers.
- Education needed for particular careers.

Have students break into small groups to share their career explorations. After sufficient time, bring the smaller groups together for a larger discussion of what students have discovered as possible careers.

This may conclude Session 5 for a two-session approach to this topic.

Ongoing for Session 5 or New Session 6

As students have now reviewed their career exploration and related education needed for particular careers, let us move to a review of early college searches.

Again, in the voice of the advisor/counselor…

Now that you have reviewed your initial research into possible careers, let's look back at what you have learned in your college search. Keep in mind that as part of career exploration, you investigated possible **majors** and how this relates to your college exploration.

Again, let's review some key terms essential to college searches:

- Liberal arts colleges.
- Universities.
- Engineering/Architecture.
- Education.
- Business.
- Science.

After review, either break students into small groups or have one larger discussion where students share their possible career and college paths. Remember to remind students that what they have researched is a first step in the all-important journey to finding an area of study and career path that they will find fulfilling.

This meeting is also a perfect time to speak to students about course selection options that they may have in high school. Include in this discussion:

- Advanced Placement classes in subject areas where student performance on the AP Exam may receive college credit at the college or university where they matriculate.
- Honors classes in subject areas where they excel are a selection for many students.

A Final Session

The aim of this session is to have students look back at their middle school journey and to some extent analyze what they have learned about themselves as a student.

Have students complete the following exercise.

EXERCISE 15.3 Self Reflection

What have I learned about myself as a student? What are my strengths? What do I need to work on in high school?
At this point, do you have any possible career paths? Any college interests?

After students complete this exercise, have either a small group or one large group discussion where students may share their thoughts.

At the conclusion of this session, advisors/counselors may want to celebrate this final session in some way: treats in the cafeteria,

Part II. Helping Parents as Their Child Transitions to High School

SAMPLE AGENDA OF A PARENTS AND EIGHTH GRADERS MEETING (to Be Held in Early/Mid May)

In this sample, both parents and eighth graders meet in the auditorium/gym to discuss the transitioning process.

Aim: This meeting has a two-fold purpose:

1. Reminding parents of the fundamental skills that their children have fine-tuned since sixth grade, and the career and college early exploration that their children have done.
2. Have students share **Exercise 15.3** with their parents.

Principal and Head of the Counseling Department welcome parents and students in an auditorium/gym setting. In this meeting, the Principal and Head of the Counseling Department review what the advisory/group counseling sessions have covered in the sixth — eighth grades with emphasis on the second semester of eighth grade. They also refer to the Parent Meetings in sixth and seventh grades that have addressed these topics.

- Fundamental study skills including planning, organization, and time management.
- Early career exploration and college searches as motivating factors.

After this presentation, have parents and child break into separate spaces so that children may speak individually with their parents using a copy of **Exercise 15.3 Self Reflection.** The purpose of this part of the meeting is to ensure that parents understand their children's perception of themselves as students as they transition to high school. (Announce the allotted time for this part of the meeting.)

After the allotted time has passed, the principal calls parents and eighth graders back to the general meeting. At this point, a question-and-answer session may follow.

Principal concludes with a brief summary of how middle school children have acquired the foundational skills needed for secondary education. Concluding remarks should include informing parents about contacting the appropriate staff member with any questions regarding their child's transition to high school.

A Final Note

Following the steps outlined in the shared agenda of the advisory/group counseling sessions over the course of middle school has prepared eighth graders with the foundational basis that is essential for success in high school and beyond. As middle schoolers enter the next stage of their K–12 journeys, secondary education, not only are they prepared to meet the challenges that they will encounter, but also to thrive in this new environment!

- See Chapter 14 for other agendas of Meetings with Parents.

Notes

1 Aromolaran, Simisola K. *Challenges of Middle School to High School Transition: A Phenomenological Study of the Experiences of Students Transitioning From Eighth Grade to Ninth Grade.* Digital Commons @ ACU, Electronic Theses and Dissertations. Paper 355.

2 Legters, Nettie and Kerri Kerr. "Easing the Transition to High School: An Investigation of Reform Practices to Promote Ninth Grade Success." *Escholarship.org*, January 2001. https://escholarship.org/uc/item/26m5j3dr

3 "What's Your Learning Style?" *EducationPlanner.Org.* https://www.educationplanner.org/students/self-assessments/learning-styles

4 Curry, Jennifer R. and Amy Milsom. *Career and College Readiness Counseling in P-12 Schools*, 2nd edition. New York: Springer Publishing Co., 2017, p. 236.
5 Curry, Jennifer R. and Amy Milsom. *Career and College Readiness Counseling in P-12 Schools*, 2nd edition. New York: Springer Publishing Co., 2017, p. 237.
6 Curry, Jennifer R. and Amy Milsom. *Career and College Readiness Counseling in P-12 Schools*, 2nd edition. New York: Springer Publishing Co., 2017, pp. 219 and 239.
7 Curry, Jennifer R. and Amy Milsom. *Career and College Readiness Counseling in P-12 Schools*, 2nd edition. New York: Springer Publishing Co., 2017, pp. 219 and 239.

16

Some Final Thoughts

Our purpose in writing *Counseling and Advising 21st Century Middle School Students: An Interactive Curriculum for Academic Growth, Career and College Readiness, and Social Emotional Development* is to emphasize the essential importance of the middle school years academically, cognitively, psychosocially, and socially–emotionally. In this last chapter, we want to first consider the developmental overview for eighth grade students who will be engaging in high school transition planning. As these students move to the next step of their K–12 journeys, hopefully what they have experienced in middle school has solidly prepared them for the increased demands of high school. Students entering ninth grade should be well prepared to actively engage in yet another new world of education. We offer a final overview that includes the developmental characteristics of our eighth graders, including academic, cognitive, psychosocial, and social–emotional growth as they leave middle school.

Initially, we want to emphasize the key importance that an advisory system or group counseling program plays in establishing the academic foundational basis that prepares middle school students for secondary education and the world of possibilities that postsecondary education provides. From establishing goals and developing effective strategies to achieve them, to the enhancement of study skills and habits, middle school students have learned how to achieve greater academic success. At

DOI: 10.4324/9781003597292-19

the same time, the early exposure to career and college exploration in the lives of these middle school students has developed a true sense of openness to both career and college possibilities. As they have explored careers and performed an initial search of colleges, they have experienced views of their potential future selves. This readiness will hopefully become a highly motivating factor in their young lives as they transition to high school and later the postsecondary world. Indeed, the active role of counselors, teachers and administrators is essential in helping middle school students establish a positive outlook as they move successfully into the more expansive world of high school.

Additionally, the active role of parents is critical in the life of a middle school student. While the classroom may appear dramatically different as students work in Google Classroom or Schoology, the building of fundamental skills now continues in a new framework that encourages a more interactive curriculum. It is essential to keep ongoing communication with parents, not only about their child's progress, but also about what is being accomplished in advisory system/group counseling sessions. Parents, too, may be entering a new world as their child enters middle school. Frequent meetings with parents are a necessity as students engage in the shared agenda of the advisory system or the group counseling program.

Academic Growth and Development

As educators, we have both experienced the excitement in the middle school classroom—excitement about new ideas, a strong, developing sense of independent thinking, and an openness to what the future may hold. Adolescents are truly at a pivotal stage in their young lives. Middle school is an essential time for educators to assist students in an approach to their studies that may hopefully create a fundamental basis built upon solid study skills and habits. This all-important foundation is the underlying aim of the College and Career Readiness Standards that are so essential for future success as students transition to high school and beyond. At the same time, this fundamental basis of a solid methodology in approaching their academic work will naturally extend into other areas of their young lives as they learn to balance the many demands of extracurricular activities, sports, and family responsibilities.

The Lemberger-Truelove et al. study clearly demonstrates the effectiveness of educators working with young adolescents to build this foundational basis by developing strong study habits and skills necessary for academic success. This study also indicates that this fundamental basis is essential not only for academic success, but also for college and career readiness as well as social and emotional development. Most importantly, this study demonstrates that middle school students are capable of understanding executive functioning and its related components: the ability to inhibit, to shift, to plan, to organize, and to monitor tasks. In other words, the capability of a student to avoid distractions, to move from one subject to another, to plan and to organize effectively, and to monitor and/or complete given assignments.[1] What the Lemberger-Truelove et al. study is defining is a student with the requisite foundational study skills and habits that are so necessary for academic achievement. Middle school students are at a prime age for this development that opens their young minds to a methodical approach to task completion. We as coauthors have experienced the positive effects of sixth graders understanding executive functioning and its resulting rise in grade achievement.

At the same time, as middle school students explore careers, they are at an opportune time to think about their possible future selves. Through the advisory/group counseling systems, sixth and seventh graders actively research both careers and colleges. These early explorations, based on the ACT's Inventories—Interests, Abilities, and Values—help students understand themselves in a new way that directs them to career areas matching their answers to the Inventories. As educators, we have experienced the enthusiastic excitement of middle school students in this research and especially at Career Fairs. Adolescents in this stage of development are open to exploring not only their future, but also the higher education needed for specific career fields. As students engage in a preliminary college search, they come to understand the pathways to future career endeavors. Given the solid foundational basis of strong study skills and habits and the ongoing enthusiasm of early career and college exploration, eighth grade students are truly prepared for the transition to the next step of their K–12 journeys that is high school.

Cognitive and Psychosocial Development

As Piaget noted, students in middle school develop more advanced thinking abilities than what they had possessed in elementary school; in other words, students are entering the stage of formal operations. As noted in Chapter 5, according to the Cleveland Clinic, brain development in adolescents allows them to "deal with abstractions, test hypotheses and see infinite possibilities...the frontal cortex continues maturing until people reach their mid to late 20s... This area of the brain controls executive functions such as planning, prioritizing and controlling impulses."[2] Developmentally, middle school students are capable of understanding and accomplishing executive functioning and its related components—effective planning, organization, time management, and avoidance of distractions. In addition, middle school students certainly should understand cause and effect, particularly as it applies to their lives in school. Furthermore, the Wigfield et al. study found that middle school students are capable of abstract thought and are engaged in the process of organization and reflection. This abstract thinking gives adolescents new skills, including thinking logically.[3] The Lemberger-Truelove et al. study further found that the teen brain is highly adaptable, allowing for strengthening of circuits through learning, exercise, and "positive" risks like joining teams, but also making it vulnerable to negative environmental impacts.[4] Furthermore, as the ABCD Study shows, adolescence is a period of dramatic brain development.[5] For the eighth graders especially, the emergence of formal operational thinking supports the further cognitive developmental ability to understand and to take on the understanding of another person's perspective. What these studies confirm is that middle school students are at a prime age to engage not only in critical thinking that includes synthesis and analysis, but also that they are ready to develop the executive functioning skills that are the foundational basis for their future success.

Educators also need to understand that middle school students are in a period of psychological development that is open to new ideas and approaches to achieving academic and personal success. According to Erik Erikson (1963), an early researcher of the theory of psychosocial development, eighth graders are engaged in the well-known *identity versus role confusion stage.* Adolescents struggle to identify who they want to be currently and in their future. Part of the challenge to develop their true identity is that they struggle to reconcile their desires with their friends and parents. These students are highly influenced by their peers.[6] Later, Marcia determined that identity and role confusion are crucial in adolescent development. Furthermore, he identified four developmental statuses: identity diffusion, identity foreclosure, identity moratorium, and identity achievement. As noted in Chapter 2, Marcia's work presents an ongoing developmental approach in his identification of the four statuses that he uses to describe how identity achievement, the final status, is realized. This development enables adolescents to explore future ideas as they further understand themselves in order to reach the identity achievement status.[7] Furthermore, social development is a major important concern for adolescence. They begin to strengthen their psychosocial development to better connect psychosocially with their peers. Finally, it is very important that these eighth graders learn how to develop positive communication and conflict resolution strategies to move forward their psychosocial development.

Social and Emotional Growth

Just as middle school students are developing cognitively and psychosocially in this crucial stage of their young lives, they are also experiencing social and emotional growth. Quite literally, they are in a process where they are striving to develop the skills, attitudes, and values necessary to acquire social and emotional competence.[8] Clearly, this period of their young lives is highly formative as they are establishing the foundation of what Albert Bandura referred to as the time in which adolescents develop a belief in their own self-efficacy. As noted in Chapter 3, Bandura's self-efficacy theory stresses that ability perception appears to have a major impact on choice behavior and defines self-efficacy as the expectation to successfully perform a behavior required for a particular task.[9] In the life of a middle schooler, this self-perception may have a powerful impact on their view of themselves not only academically but also socially and emotionally.

Furthermore, as noted above, the Lemberger-Truelove et al. study also demonstrated a positive outcome in early counselor intervention in working with middle school students. Similarly, the Wigfield et al. study also promotes active counselor involvement with students and parents alike in the transition from elementary school to middle school. Indeed, David Yeager further advocates for a new model for social and emotional learning environment that is based upon a positive climate and healthy mindsets during these highly formative middle school years.[10] At the same time, some of the middle school students are experiencing anxiety, depression, self-harm, and suicidal thoughts. Former Surgeon General Vivek Murthy brings particular attention to this subject. Issues presented in *Murthy's Letter* call for increased school counselors and mental health counselors to help middle school students struggling with these issues.[11]

Teachers, counselors, and administrators play such important roles in the lives of middle school students. None of this foundational building would be possible without these educators who are so committed to helping middle school students not only succeed but also thrive as they move into high school. Together, these educators have a strong, positive impact on middle school students, helping them as students mature and understand themselves personally and academically. With this self-knowledge, students will begin to formulate their future aspirations. At the same time, as school budgets continue to tighten, we are so thankful for the essential role that teachers, counselors, and administrators play as advocates for their students. We urge teachers, counselors, and administrators to be increasingly proactive on behalf of their students.

We as coauthors have had extensive experience in working in 6–12 schools. As noted in Chapter 3, we have addressed multiple educational issues in our prior book, *Counseling 21st Century Students for Optimal College and Career Readiness, A 9th-12th Grade Curriculum* Second Edition. Now as coauthors, we have been able to speak about the many personal issues that a middle school student encounters in this all-important developmental period. We understand that counselors, teachers, and administrators have always played a critical role in the development of their students. As the world becomes more complex and as students are increasingly exposed to many forces during their developmental years, our many roles take on even more importance. Both of us have experienced the roles of counselor, teacher, and administrator; this collective experience has broadened our perspectives of what adolescents need over time for successful growth and development. Our experience with middle school students has only reenforced our long-held belief that helping students involves the work of all—teachers, counselors, administrators, parents, and students alike—as adolescents successfully navigate their ways through middle school. We also recognize the difficulties that many middle school students may experience. At the same time, we remain hopeful that middle school will prepare students for the next step of their K–12 journeys as they successfully transition to high school.

A world of possibilities awaits our students in the postsecondary stage of their lives. Whether they attend an apprenticeship program, community college, or a four-year college and beyond, what middle school students have learned in this crucial time period provides the foundational basis not only for high school, but also for postsecondary options that may hopefully lead to a rewarding career and life experience. Kudos to all!

Notes

1 Lemberger-Truelove, Matthew et al. "Growth in Middle School Student's Curiosity, Executive Functioning and Academic Achievement." *Professional School Counseling*, May 2021, Vol. 24, No. 1b, Special Issue: Educator–Counselor: A Nondual Identity for School Counselors, p. 1.

2 "Adolescent Development." *Cleveland Clinic.* https://my.clevelandclinic.org/health/articles/7060-adolescent-development
3 Wigfield, Alan, et al. "Early Adolescents' Development Across the Middle School Years: Implications for School Counselors." *Professional School Counseling*, December 2005, Vol. 9, No. 2, Special Issue: Focus On Middle School Counseling, p. 113.
4 Lemberger-Truelove, Matthew et al. "Growth in Middle School Student's Curiosity, Executive Functioning and Academic Achievement." *Professional School Counseling*, May 2021, Vol. 24, No. 1b, Special Issue: Educator–Counselor: A Nondual Identity for School Counselors.
5 "Adolescent Brain Cognitive Development Study (ABCD Study)." *National Institute of Mental Health (NIMH).* https://www.nimh.nih.gov/research/research-funded-by-nimh/research-initiatives/adolescent-brain-cognitive-developmentsm-study-abcd-studyr
6 Erikson, E. H. *Childhood and Society*, 2nd edition. New York: W. W. Norton & Co., 1993.
7 Marcia, James, et al. *Ego Identity: A Handbook for Psychosocial Research*, 2nd edition. New York: Springer-Verlag, 1993.
8 Taylor, Howard E. and Susan Larson. "Social and Emotional Learning in Middle School." *The Clearing House*, Jul–August, 1999, Vol. 72, No. 6, p. 331.
9 Bandura, Albert. *Social Foundation of Thought and Action: A Social Cognitive Theory.* Prentice Hall, Inc, 1986.
10 Yeager, David S. "Social and Emotional Learning Programs for Adolescents." *The Future of Children*, Spring 2017, Vol. 27, No. 1, Social and Emotional Learning, Princeton University, pp. 73–94.
11 Richtel, Matt. "The Surgeon General's New Mission: Adolescent Mental Health." *New York Times*, March 21, 2023. https://www.nytimes.com/2023/03/21/health/surgeon-general-adolescents-mental-health.html

Appendix A: ACT Resources

https://www.act.org/

ACT Resources—The ACT Test–K–12 Administrators | ACT

https://www.act.org/content/act/en/k12-educators-and-administrators/administrator-resources/the-act.html

Resources to help students and parents navigate the ACT National Test and ACT District Testing.

K–12 Toolkits–K–12 Administrator Resources | ACT

https://www.act.org/content/act/en/k12-educators-and-administrators/administrator-resources/toolkits.html

Communicate and engage students, parents, and educators about the ACT college and career readiness programs you offer.

ACT Career Readiness Resources | K–12 Administrators | ACT

https://www.act.org/content/act/en/k12-educators-and-administrators.html

Career readiness resources for K–12 professionals to help make your students unstoppable. Track students' progress and prepare them for success through high school and beyond.

ACT State and District Testing | K–12 Resources for Administrators | ACT

https://www.act.org/content/act/en/products-and-services/the-act-educator/district-testing-administrators.html

Information for K–12 Administrators about the benefits of offering ACT standardized testing on a school day. Learn more about how to leverage federal funding with our free toolkit!

ACT Resources—Scores and Reporting—K–12 Administrators | ACT

https://www.act.org/content/act/en/k12-educators-and-administrators/administrator-resources/scores-and-reporting.html

Understand the data that you receive from student score reports and how to use it to improve curriculum and student success outcomes.

ACT Resources—Social Emotional Learning—K–12 Administrators | ACT

https://www.act.org/content/act/en/k12-educators-and-administrators/administrator-resources/social-emotional-learning.html

Social emotional learning resources for K–12 professionals to help make your students unstoppable.

School Counselors' Perspectives on the Social and Emotional Development of Students—ACT Research

https://www.act.org/content/act/en/research/reports/act-publications/school-counselors-perspectives-on-the-social-and-emotional-development-of-students.html

ACT collaborated with the American School Counselor Association (ASCA) to identify school counselor perspectives, needs, and challenges related to students' social and emotional development.

Appendix B: College Board Resources

College Board K–12 Educators Brief

The College and Career Readiness Benchmarks for the SAT Suite of Assessments

https://satsuite.collegeboard.org/media/pdf/educator-benchmark-brief.pdf

The new benchmarks differ from the old ones in several key ways:

1. Focus on Section-Level Metrics:
   - Previously, the SAT had a single college readiness benchmark of 1550 for the combined scores of three sections (Critical Reading, Mathematics, and Writing). This indicated a 65% probability of achieving a B– or higher in the first year of college.
   - The new benchmarks are set at the section level (Math and Evidence-Based Reading and Writing) and represent a 75% likelihood of achieving at least a C grade in first-semester, credit-bearing college courses in related subjects.

2. Subject-Specific Readiness:
   - The old benchmarks provided a general sense of college readiness but did not identify the types of courses students might be ready for or the skills they had developed.
   - The new benchmarks are designed to reflect readiness for specific first-semester college courses, such as algebra, statistics, precalculus, calculus, history, literature, social science, or writing.

3. Grade-Level Benchmarks:
   - The new benchmarks include grade-level benchmarks for the PSAT/NMSQT, PSAT 10, and PSAT 8/9, which are based on expected student growth toward the SAT benchmarks at each grade level. This allows tracking progress year-over-year.

SAT® Score Relationships with College GPA: First-Year through Fourth-Year Cumulative GPA

https://research.collegeboard.org/media/pdf/SAT%28R%29%20Score%20Relationships%20with%20College%20GPA.pdf

Jessica P. Marini, Paul A. Westrick, Linda Young, and Emily J. Shaw October 2024

Key Takeaways

- SAT scores remain consistently predictive of cumulative GPA throughout each year of college -- —from first-year GPA through fourth-year GPA.
- For all subgroups of students and institutions studied, including underrepresented minority students, students from high-challenge environments, first-generation college students, and students whose best language is not English, SAT scores retain their predictive power through each year of college.
- The information added by SAT scores above HSGPA to predict college GPA remains consistent through the first four years of college. This was true for all subgroups examined, as well.
- Colleges and universities that use SAT scores for admissions, placement, scholarships, and academic advising can be confident that test scores predict how students are expected to perform throughout college.

SAT Suite of Assessments—Educator Guide

The SAT Suite of Assessments Educator Guide connects schools directly with our online resources. Schools won't receive a print or PDF version of the guide. https://satsuite.collegeboard.org/sat-suite-educator-guide

If you're an educator, counselor, or caring adult, watch this video to learn how you can use Career Insights Snapshot to support your student's career journey.

https://bigfuture.collegeboard.org/help-center/what-resources-are-available-educators-counselors-and-other-caring-adults-help-better

BigFuture for Educators—BigFuture

Educator resources to help students after high school including college search, career search, and more.

https://bigfuture.collegeboard.org/communities-events/for-educators

Index

For Product Safety Concerns and Information please contact our EU representative GPSR@taylorandfrancis.com Taylor & Francis Verlag GmbH, Kaufingerstraße 24, 80331 München, Germany